Node.js 12 实战

赵荣娇 著

清华大学出版社
北京

内 容 简 介

本书以Node.js 12版本为基础,以代码演练为讲解方式,介绍Node.js开发中涉及的技术。本书简单实用,可以从零学起,方便初学者快速入门。

本书共12章,涵盖的主要内容有:Node.js与JavaScript的联系和区别、Node.js工作原理和NPM的使用、Node.js语法和常用模块、Node.js异步I/O与多线程、Node.js错误处理、Node.js测试方法、Node.js的数据处理方法、结合Vue+Express+Koa+MongoDB的Node.js项目实战等。

本书内容丰富,实例典型,实用性强,适合有一定的HTML、CSS、JavaScript基础,希望全面学习Node.js开发的前端开发人员阅读,也适合希望提高Web项目开发水平的人员阅读。

本书封面贴有清华大学出版社防伪标签,无标签者不得销售。
版权所有,侵权必究。侵权举报电话:010-62782989　13701121933

图书在版编目(CIP)数据

Node.js 12实战/赵荣娇著.— 北京:清华大学出版社,2020.7
ISBN 978-7-302-55706-7

Ⅰ.①N… Ⅱ.①赵… Ⅲ.①JAVA语言－程序设计 Ⅳ.①TP312.8

中国版本图书馆CIP数据核字(2020)第105016号

责任编辑:夏毓彦
封面设计:王　翔
责任校对:闫秀华
责任印制:宋　林

出版发行:清华大学出版社
　　　　网　　址:http://www.tup.com.cn,http://www.wqbook.com
　　　　地　　址:北京清华大学学研大厦A座　　邮　　编:100084
　　　　社 总 机:010-62770175　　　　　　　　邮　　购:010-62786544
　　　　投稿与读者服务:010-62776969,c-service@tup.tsinghua.edu.cn
　　　　质量反馈:010-62772015,zhiliang@tup.tsinghua.edu.cn
印 装 者:三河市龙大印装有限公司
经　　销:全国新华书店
开　　本:190mm×260mm　　印　张:21.5　　字　数:551千字
版　　次:2020年8月第1版　　　　　　　　印　次:2020年8月第1次印刷
定　　价:69.00元

产品编号:087547-01

前　言

　　Node.js 是一个基于 Chrome V8 引擎的 JavaScript 运行环境，它也是一个让 JavaScript 运行在服务端的开发平台。它让 JavaScript 成为与 PHP、Python、Perl、Ruby 等服务端语言平起平坐的脚本语言。

　　目前图书市场上关于 Node.js 开发及实践的图书不少，本书吸取已出版同类图书的优点，从实际应用出发，通过各种典型模块和项目案例来指导读者提高 Web 应用开发水平。本书以实战为主旨，通过 Node.js 开发中涉及的基础知识和 3 个完整的项目案例，让读者全面、深入、透彻地理解 Node.js 开发技术栈的整合使用（如 Vue+Express、Koa+MongoDB 等），提高实际开发水平和项目实战能力。

本书特色

1. 附带源码，提高学习效率

　　为了便于读者理解本书内容，提高学习效率，作者专门为本书每一章内容都附上所需的实战源代码，可下载使用。

2. 涵盖 Node.js 开发的各种热门技术及其整合使用

　　本书涵盖 NPM、ES6、常用模块，包括文件系统、HTTP、TCP、WebSocket、Events 等热门技术及整个技术栈框架的整合使用。

3. 对 Node.js 开发的各种技术和框架进行原理上的分析

　　本书从一开始便对 Web 开发基础和 Node.js 开发的环境配置做了基本介绍，并对各种开发技术及其整合进行了原理性的分析，便于读者理解书中的典型模块开发和项目案例。

4. 项目案例典型，实战性强，有较高的应用价值

　　本书最后提供了 3 个项目实战案例。这些案例来源于作者所开发的实际项目，具有很高的应用价值和参考性，便于读者融会贯通地理解本书中所介绍的 Node.js 技术。读者将案例稍加修改，便可用于实际项目开发中。

源代码下载

本书配套示例源代码可通过扫描下边的二维码下载。

如果下载有问题，请联系 booksaga@163.com，邮件主题为"Node.js 12 实战"。

本书读者

- 希望全面学习 Node.js 开发的 Web 前端开发人员。
- 希望提高项目开发水平的前端开发人员。
- IT 技术培训机构的师生。
- 需要一本 Node.js 开发案头必备查询手册的人员。

作者简介

赵荣娇，飞猪旅行前端开发工程师，擅长 CSS、JavaScript 和各种框架，参与写作或翻译过多本前端精品图书。喜欢旅行，热爱前端开发，乐于分享。

作　者
2020 年 3 月

目　录

第 1 章　Node.js 与 JavaScript·······1

- 1.1　JavaScript 与前端·······1
- 1.2　第一个 JavaScript 实现的 Hello World·······1
- 1.3　Node.js 12 安装前的准备·······3
  - 1.3.1　在 Microsoft Windows 系统上安装 Node.js·······3
  - 1.3.2　在 Linux 发行版上安装 Node.js·······5
  - 1.3.3　在 Mac OS X 上安装 Node.js·······7
  - 1.3.4　Mac OS X 中 Node.js 版本的切换和升级·······7
- 1.4　JavaScript 与 Node.js 对比·······10
- 1.5　第一个 Node.js 实现的 Hello World·······10
  - 1.5.1　纯脚本示例·······10
  - 1.5.2　交互模式·······11
  - 1.5.3　创建 Node.js 项目·······11
- 1.6　Node.js 的开发调试工具 Inspect·······12

第 2 章　NPM、REPL 与 Node.js 工作原理·······15

- 2.1　NPM 的使用·······15
  - 2.1.1　NPM 简介·······15
  - 2.1.2　NPM 的工作原理·······16
  - 2.1.3　package.json 属性说明·······17
  - 2.1.4　NPM 的常用命令·······19
- 2.2　REPL 的使用·······21
- 2.3　Runtime 和 vm·······22
  - 2.3.1　Runtime·······22
  - 2.3.2　vm·······23

2.4 回调函数 ·· 24
2.5 同步/异步和阻塞/非阻塞 ·· 25
  2.5.1 同步和异步 ··· 26
  2.5.2 阻塞和非阻塞 ·· 28
  2.5.3 同步/异步和阻塞/非阻塞 ··· 29
2.6 单线程和多线程 ·· 29
2.7 并行和并发 ·· 30
2.8 事件循环 ··· 31

第 3 章 Node.js 的语法 ·· 34

3.1 ECMAScript 6 标准 ·· 34
3.2 数组常用方法及 ES6 中的数组方法 ··· 34
3.3 函数 ··· 40
  3.3.1 参数的默认值 ·· 40
  3.3.2 rest 参数 ·· 44
  3.3.3 name 属性 ·· 45
  3.3.4 箭头函数 ·· 46
3.4 闭包 ··· 49
3.5 对象 ··· 51
  3.5.1 属性的简洁表示 ··· 52
  3.5.2 属性名表达式 ·· 53
  3.5.3 方法的 name 属性 ·· 55
  3.5.4 对象的扩展运算符 ·· 56
  3.5.5 对象的新方法 ·· 58
  3.5.6 属性的可枚举性 ··· 67
  3.5.7 属性的遍历 ··· 68
3.6 类 ·· 69
  3.6.1 基础用法 ·· 69
  3.6.2 封装与继承 ··· 74
  3.6.3 super 关键字 ··· 77
3.7 ES6 的模块化 ··· 78
  3.7.1 基本用法 ·· 78

3.7.2　as 的用法 79
　　　3.7.3　import 命令的特点 80
　　　3.7.4　export 与 import 81
　3.8　使用 Babel 转码 82
　3.9　使用 N-API 84

第 4 章　Node.js 常用模块 88

　4.1　Module 88
　　　4.1.1　创建和使用模块 88
　　　4.1.2　require 方法中的文件查找策略 89
　4.2　Buffer 93
　　　4.2.1　Buffer 与字符编码及转换 93
　　　4.2.2　Buffer 类及其方法 94
　　　4.2.3　Buffer 与性能 99
　4.3　File System 99
　　　4.3.1　异步读文件 100
　　　4.3.2　同步读文件 101
　　　4.3.3　打开文件 102
　　　4.3.4　写入文件 103
　　　4.3.5　获取文件信息 104
　　　4.3.6　fs.read 异步读文件 105
　　　4.3.7　fs.close 异步关闭文件 106
　4.4　HTTP/HTTP2 服务 107
　　　4.4.1　http 模块 108
　　　4.4.2　http2 模块 111
　4.5　TCP 服务 114
　　　4.5.1　构建 TCP 服务器 115
　　　4.5.2　服务器和客户端之间的通信 118
　　　4.5.3　构建 TCP 客户端 119
　4.6　SSL 121
　　　4.6.1　SSL 简介 121
　　　4.6.2　使用 OpenSSL 进行证书生成 123

   4.6.3 Node.js 实现 HTTPS 的配置 ··· 125
 4.7 WebSocket ··· 126
   4.7.1 ws 模块 ·· 126
   4.7.2 实战：ws 简易聊天室 ··· 128
 4.8 流 ·· 136
   4.8.1 可读流 ·· 137
   4.8.2 可写流 ·· 138
   4.8.3 管道流 ·· 140
 4.9 事件 ·· 141
   4.9.1 注册事件名&监听器 ·· 142
   4.9.2 给监听器 listener 传入参数与 this ································ 143
   4.9.3 最多只触发一次的监听器 ··· 143
   4.9.4 添加监听器/移除监听器事件 ··· 144
   4.9.5 错误事件 ·· 144
 4.10 实战演练 RESTful API ·· 145

第 5 章 Node.js 调试 ··· 152

 5.1 基础调试 ··· 152
   5.1.1 基础 API ·· 152
   5.1.2 自定义 stdout ··· 154
   5.1.3 控制调试日志 ·· 155
 5.2 进阶调试 ··· 160
   5.2.1 使用 Inspect 调试 ·· 161
   5.2.2 使用 VSCode IDE 调试 ··· 164

第 6 章 Node.js 的异步 I/O 与多线程 ···························· 167

 6.1 异步 I/O ·· 167
   6.1.1 异步 I/O 的必要性 ··· 167
   6.1.2 操作系统对异步 I/O 的支持 ·· 168
   6.1.3 异步 I/O 与轮询技术 ·· 169
 6.2 进程、线程、协程等 ··· 169
   6.2.1 进程、线程、协程 ·· 169
   6.2.2 应用场景 ·· 171

6.2.3　并发与并行 171

6.3　在 Node.js 中实现多线程 172

　　6.3.1　单线程的 JavaScript 172

　　6.3.2　Node.js 内部分层 174

　　6.3.3　libuv 174

　　6.3.4　多进程 175

6.4　Node 性能小结 176

第 7 章　Node.js 的错误处理 177

7.1　错误的分类 177

7.2　函数的错误处理 178

7.3　实战演练异常-错误处理 179

　　7.3.1　同步代码的异常捕获处理 179

　　7.3.2　异步代码的错误处理 180

　　7.3.3　使用 event 方式来处理异常 181

　　7.3.4　Callback 方式 182

　　7.3.5　Promise 方式 182

　　7.3.6　使用 domain 模块 185

　　7.3.7　多进程模式加异常捕获后重启 186

第 8 章　Node.js 的测试 188

8.1　什么是单元测试 188

8.2　一个简单的单元测试 189

8.3　Mocha 190

8.4　Assert 192

8.5　测试 HTTP 接口 196

8.6　代码覆盖率工具 istanbul 197

第 9 章　Node.js 的数据处理 199

9.1　MySQL 199

　　9.1.1　Node.js 连接 MySQL 199

　　9.1.2　数据库操作 200

　　9.1.3　使用 Sequelize 操作数据库 202

9.2 MongoDB ·· 208
    9.2.1 创建数据库 ·· 208
    9.2.2 数据库操作 ·· 209
9.3 Redis ·· 215
    9.3.1 Node.js 连接 Redis ·· 215
    9.3.2 列表——List ··· 217
    9.3.3 集合——Set ·· 218
    9.3.4 消息中介 ·· 218

第 10 章 实战：使用原生 JavaScript 开发 Node.js 案例 ············· 220

10.1 项目任务 ·· 220
10.2 HTTP 服务器 ·· 221
10.3 服务端模块化 ·· 223
10.4 设计请求路由 ·· 224
10.5 请求处理程序 ·· 229
10.6 非阻塞式处理请求响应 ·· 232
10.7 处理 POST 请求 ·· 235
10.8 文件上传 ·· 241

第 11 章 实战：基于 Vue+Express+MongoDB 实现一个后台管理系统 ············· 248

11.1 项目结构 ·· 248
11.2 前端代码实现 ·· 249
    11.2.1 项目依赖的模块 ·· 249
    11.2.2 注册页 ·· 250
    11.2.3 登录页 ·· 254
    11.2.4 管理页 ·· 258
11.3 后端代码实现 ·· 266
    11.3.1 数据库设计 ·· 266
    11.3.2 启动应用 ··· 268
    11.3.3 注册/登录接口 ·· 268
    11.3.4 增删改查接口 ··· 271

第 12 章　实战：基于 Koa+MongoDB 实现博客网站 …… 279

12.1　项目结构 …… 279
12.2　数据库设计 …… 281
12.2.1　数据准备 …… 281
12.2.2　连接数据库 …… 281
12.2.3　创建表结构 …… 282
12.3　服务端实现 …… 283
12.3.1　启动后台应用 …… 283
12.3.2　配置中间件 …… 283
12.3.3　搭建路由和控制器 …… 287
12.3.4　账户管理 …… 290
12.3.5　博客管理 …… 294
12.4　博客后台管理的实现 …… 297
12.4.1　目录结构 …… 297
12.4.2　权限管理 …… 297
12.4.3　博客管理 …… 309
12.5　博客前台站点的实现 …… 322
12.5.1　目录结构 …… 322
12.5.2　博客列表页 …… 323
12.5.3　博客详情页 …… 327

第 1 章
Node.js与JavaScript

Node.js 是运行在服务端的 JavaScript，是一个基于 Chrome V8 引擎的 JavaScript 运行环境。换句话说，Node.js 是一款工具，是一个基于 Chrome V8 引擎的在服务端运行 JavaScript 代码的工具。Node.js 为 JavaScript 在服务端运行提供了一个运行环境。

1.1 JavaScript 与前端

JavaScript 是一种 Web 编程语言。

截至 2012 年，所有浏览器都完整地支持 ECMAScript 5.1，旧版本的浏览器至少支持 ECMAScript 3 标准。1996 年 11 月，JavaScript 的创造者 Netscape 公司决定将 JavaScript 提交给国际标准化组织 ECMA，希望这门语言能够成为国际标准。次年，ECMA 发布 262 号标准文件（ECMA-262）的第一版，规定了浏览器脚本语言的标准，并将这种语言称为 ECMAScript，这个版本就是 1.0 版。2015 年 6 月 17 日，ECMA 国际组织发布了 ECMAScript 的第 6 版，该版本正式名称为 ECMAScript 2015，但通常被称为 ECMAScript 6 或者 ES6。

ECMAScript 标准一开始就是针对 JavaScript 语言制定的，但是没有称其为 JavaScript，有两个方面的原因：一是商标，JavaScript 本身已被 Netscape 注册为商标；二是想体现这门语言的制定者是 ECMA，而不是 Netscape，这样有利于保证这门语言的开发性和中立性。

因此，ECMAScript 和 JavaScript 的关系是，前者是后者的规格，后者是前者的一种实现。尽管 ECMAScript 是一个重要的标准，但它并不是 JavaScript 唯一的部分，当然也不是唯一被标准化的部分。实际上，一个完整的 JavaScript 实现是由以下 3 个不同部分组成的：

- 核心（ECMAScript），描述了 JavaScript 语言本身的相关内容。
- 文档对象模型 DOM（Document Object Model），整合 JavaScript、CSS 和 HTML。
- 浏览器对象模型 BOM（Browser Object Model），整合 JavaScript 和浏览器。

1.2 第一个 JavaScript 实现的 Hello World

JavaScript 是一种轻量级的编程语言。本节编写第一个 JavaScript 程序，让读者有个直观且简单的了解。

【示例1-1】直接打开 Chrome 浏览器控制台，输入如下代码：

```
document.write("Hello World!");
```

在浏览器窗口可以看到有"Hello World！"字样直接显示在屏幕上，如图1.1所示。或者输入如下代码：

```
console.log ("Hello World!");
```

图1.1 通过控制台输出"Hello World！"

【示例1-2】交互式输出 Hello World 的方法。新建文件 HelloWorld.html，输入如下代码并保存文件：

```html
<!DOCTYPE html>
<html>
<head>
<meta charset="utf-8">
<title>HelloWorld</title>
<script>
function display(){
 document.getElementById("demo").innerHTML='Hello World';
}
</script>
</head>
<body>

<h1>Hello World</h1>

<button type="button" onclick="display()">显示</button>

</body>
</html>
```

使用浏览器打开该文件，并单击屏幕上的显示按钮，则会看到屏幕上显示出"Hello World"字样，如图1.2所示。

图1.2　第一个 JavaScript 实现的 Hello World

1.3　Node.js 12 安装前的准备

学习 Node.js 首先需要准备开发环境。本节介绍如何在 Windows、Linux 和 Mac OS X 上安装 Node.js。

1.3.1　在 Microsoft Windows 系统上安装 Node.js

1. 下载安装包

访问地址 https://nodejs.org/en/download/，选择 LTS 版本，如图1.3所示。选择偶数的 Node.js 版本，例如 8、10，因为 Node.js 版本迭代较快，偶数是长期稳定版本，有较好的兼容性；而奇数版本是偶数稳定版本前的开发版本。

提　示
Node.js 也常称作 NodeJS。

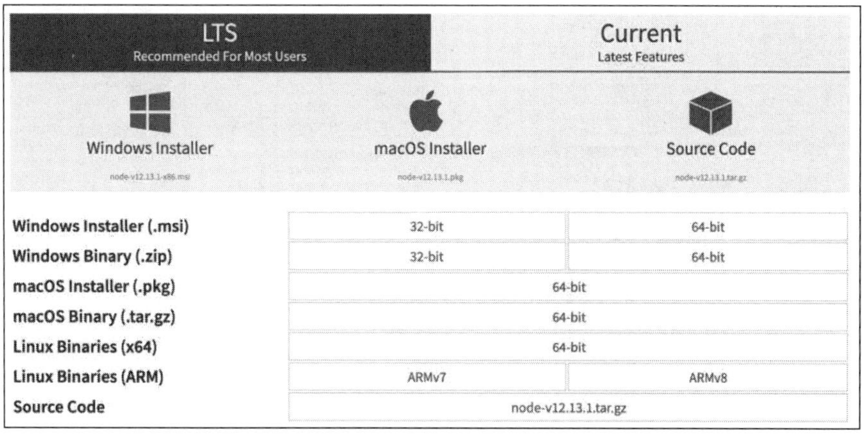

图1.3　下载 Windows 对应的安装包

2. 安装

（1）双击下载的安装包，如图 1.4 所示。

（2）勾选 I accept the terms inthe License Agreement，单击 Next（下一步）按钮，如图 1.5 所示。

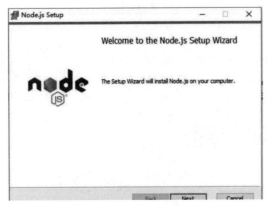
图 1.4　双击安装包

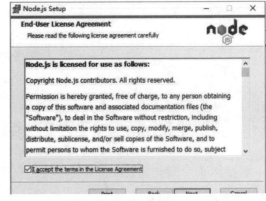
图 1.5　同意协议

（3）选择安装路径，Node.js 默认安装路径为"C:\Program Files\nodejs\"，可以修改为 D 盘，如图 1.6 所示。

（4）单击树形图标来选择你需要的安装模式，然后单击 Next 按钮，如图 1.7 所示。

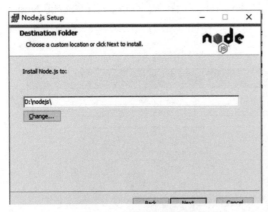
图 1.6　选择安装路径

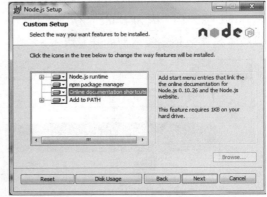
图 1.7　选择安装模式

（5）单击 Install（安装）按钮，开始安装 Node.js，如图 1.8 所示。也可以单击 Back（返回）按钮修改之前的配置项。然后单击 Next 按钮。

（6）安装过程如图 1.9 所示。

第 1 章　Node.js 与 JavaScript

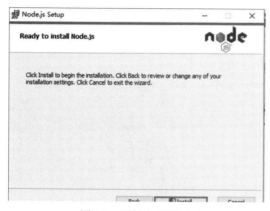

图 1.8　开始安装 Node.js

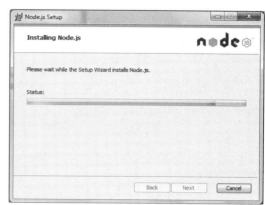

图 1.9　Node.js 安装过程

（7）单击 Finish（完成）按钮退出安装向导，如图 1.10 所示。

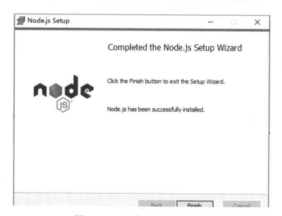

图 1.10　完成 Node.js 安装

（8）输入 node -v 查看 Node.js 版本号：

```
$ node -v
$ 12.13.1
```

新版的 Node.js 已自带 NPM，安装 Node.js 时会同时安装上。通过 NPM 可以对 Node.js 依赖的包进行管理，用于安装、卸载、更新使用 Node.js 时所依赖的包。

1.3.2　在 Linux 发行版上安装 Node.js

1. 直接使用已编译好的包

Node.js 官网上已经把 Linux 下载版本更改为已编译好的版本了，可以直接下载，解压后使用即可：

```
#   wget    https://nodejs.org/dist/v12.13.1/node-v12.13.1-linux-x64.tar.xz
// 下载
# tar xf node-v12.13.1-linux-x64.tar.xz                              // 解压
```

```
# cd node-v12.13.1-linux-x64/                          // 进入解压目录
# ./bin/node -v
12.13.1
```

解压文件，可以看到 bin 目录包含 node、npm 等命令，使用 ln 命令来设置软链接：

```
ln -s /usr/software/nodejs/bin/npm   /usr/local/bin/
ln -s /usr/software/nodejs/bin/node  /usr/local/bin/
```

2. 在 Ubuntu Linux 下使用源码安装 Node.js

以下部分将介绍如何在 Ubuntu Linux 下使用源码安装 Node.js。

（1）在 GitHub 上获取 Node.js 源码：

```
$ sudo git clone https://github.com/nodejs/node.git
Cloning into 'node'...
```

（2）修改目录权限：

```
$ sudo chmod -R 755 node
```

（3）使用 ./configure 创建编译文件：

```
$ cd node
$ sudo ./configure
$ sudo make
$ sudo make install
```

（4）查看 Node.js 的版本：

```
$ node --version
v12.13.1
```

3. 在 Ubuntu 下使用 apt-get 命令安装 Node.js

命令格式如下：

```
sudo apt-get install nodejs
sudo apt-get install npm
```

4. 在 CentOS 下使用源码安装 Node.js

（1）在 https://nodejs.org/en/download/ 下载最新的 Node.js 版本，以 v12.13.1 为例：

```
cd /usr/local/src/
wget http://nodejs.org/dist/v12.13.1/node-v12.13.1.tar.gz
```

（2）解压源码：

```
tar zxvf node-v12.13.1.tar.gz
```

（3）编译安装：

```
cd node-v12.13.1
./configure --prefix=/usr/local/node/12.13.1
make
make install
```

（4）配置 NODE_HOME，进入 profile 编辑环境变量：

```
vim /etc/profile
```

（5）设置 Node.js 环境变量，在 export PATH USER LOGNAME MAIL HOSTNAME HISTSIZE HISTCONTROL 一行的上面添加如下内容：

```
#set for nodejs
export NODE_HOME=/usr/local/node/12.13.1
export PATH=$NODE_HOME/bin:$PATH
```

（6）输入:wq 保存并退出，编译/etc/profile 使配置生效：

```
source /etc/profile
```

（7）查看 Node.js 版本以验证是否安装配置成功：

```
node -v
```

若输出"v12.13.1"字样，则表示配置成功。
npm 模块安装路径：

```
/usr/local/node/12.13.1/lib/node_modules/
```

当然，Node.js 官网提供了编译好的 Linux 二进制包，也可以下载下来直接使用。

1.3.3　在 Mac OS X 上安装 Node.js

可以通过以下两种方式在 Mac OS 上安装 Node.js：

（1）在官方网站下载 PKG 安装包，直接安装即可。
（2）使用 brew 命令来安装：

```
brew install node
```

1.3.4　Mac OS X 中 Node.js 版本的切换和升级

当开发多个项目的时候，因为不同项目所需要支持的 Node.js 版本不同，在开发过程中切换开发的项目时，经常需要切换 Node.js 版本。通过模块 n 可以管理不同版本的 Node.js。

（1）首先，使用 npm 全局安装模块 n：

```
npm install -g n
```

（2）使用命令 n 与版本号可以安装不同版本的 Node.js：

```
n 6.6.0
```

若安装成功，则显示信息如图 1.11 所示。

图 1.11　使用命令 n 安装 Node.js

（3）再使用命令 n，通过上下键选择需要用的 Node.js 版本，选择后按回车键即可，如图 1.12 所示。

图 1.12　使用命令 n 切换 Node.js 版本

（4）常用的 n 命令执行语句示例：

```
# 安装最新版本
n latest
# 安装稳定版本
n stable
# 删除某个版本
n rm x.x.x
```

使用 n -h 命令查看其所有方法：

```
n -h

  Usage: n [options/env] [COMMAND] [args]

  Environments:
    n [COMMAND] [args]            Uses default env (node)
    n io [COMMAND]                Sets env as io
    n project [COMMAND]           Uses custom env-variables to use non-official
sources
```

```
Commands:

  n                          Output versions installed
  n latest                   Install or activate the latest node release
  n -a x86 latest            As above but force 32 bit architecture
  n stable                   Install or activate the latest stable node release
  n lts                      Install or activate the latest LTS node release
  n <version>                Install node <version>
  n use <version> [args ...] Execute node <version> with [args ...]
  n bin <version>            Output bin path for <version>
  n rm <version ...>         Remove the given version(s)
  n prune                    Remove all versions except the current version
  n --latest                 Output the latest node version available
  n --stable                 Output the latest stable node version available
  n --lts                    Output the latest LTS node version available
  n ls                       Output the versions of node available

(iojs):
  n io latest                Install or activate the latest iojs release
  n io -a x86 latest         As above but force 32 bit architecture
  n io <version>             Install iojs <version>
  n io use <version> [args ...] Execute iojs <version> with [args ...]
  n io bin <version>         Output bin path for <version>
  n io rm <version ...>      Remove the given version(s)
  n io --latest              Output the latest iojs version available
  n io ls                    Output the versions of iojs available

Options:

  -V, --version   Output current version of n
  -h, --help      Display help information
  -q, --quiet     Disable curl output (if available)
  -d, --download  Download only
  -a, --arch      Override system architecture

Aliases:

  which   bin
  use     as
  list    ls
  -       rm
```

1.4 JavaScript 与 Node.js 对比

1.1 节介绍过，前端的 JavaScript 其实是由 ECMAScript、DOM、BOM 组合而成的。那么 Node.js 是由哪些部分组成的呢？

Node.js 是由以下 5 个不同部分组成的：

- ECMAScript：语言基础，如语法、数据类型结构以及一些内置对象。
- OS：即操作系统交互部分。
- File：文件系统。
- Net：网络系统。
- Database：数据库。

很容易看出，前端和后端的 JavaScript 相同点就是，二者的语言基础都是 ECMAScript，只是所扩展出来的内容不同。前端需要操作页面元素，因此需要 DOM；需要操作浏览器，因此需要扩展 BOM。而服务端的 JavaScript 则是基于 ECMAScript 扩展出了服务端所需要的一些 API，后台需要有操作系统的能力，因此扩展 OS；需要有操作文件的能力，因此扩展出 File 文件系统；需要操作网络，因此扩展出 Net 网络系统；需要操作数据，因此扩展出 Database 的能力。

前端和服务端的 JavaScript 相似，基础是相同的，但由于运行环境不同，导致所扩展出来的相关内容不同。JavaScript 是一门完整的语言，可以使用在不同的上下文中。Node.js 就是另外一种上下文，它允许脱离浏览器环境运行 JavaScript 代码，使用 V8 虚拟机来解释和执行 JavaScript 代码。同时，Node.js 提供许多有用的模块，用于简化重复工作。

因此，Node.js 事实上既是一个运行时环境，又是一个库。

1.5 第一个 Node.js 实现的 Hello World

前面理论部分讲完，本节就正式开始第一个 Node.js 项目演示。

1.5.1 纯脚本示例

【示例 1-3】首先，通过脚本模式编写第一个程序，为了编写直接输出 "Hello World" 的示例，输入：

```
console.log("Hello World!");
```

保存该文件，文件名为 helloworld.js，并通过 Node.js 命令 node 来执行该文件：

```
node helloworld.js
```

程序执行后，执行正常会在终端输出"Hello World!"：

```
node helloworld.js

Hello World!
```

1.5.2 交互模式

接下来介绍如何使用交互模式。

【示例 1-4】打开终端，输入 node 命令，进入命令交互模式，可以输入一条代码语句后立即执行并显示结果，例如：

```
$ node
> console.log('Hello World!');
Hello World!
```

效果如图 1.13 所示。

图 1.13 在交互模式下运行 Node.js 代码

1.5.3 创建 Node.js 项目

【示例 1-5】创建文件夹 test，在该目录中创建项目文件 index.js：

```
mkdir test
touch index.js
vim index.js
```

按 I 键切换至编辑模式，并将以下内容输入 index.js 文件中：

```
const http = require('http');
const hostname = '0.0.0.0';
const port = 6001;
const server = http.createServer((req, res) => {
 res.statusCode = 200;
 res.setHeader('Content-Type', 'text/plain');
 res.end('Hello World\n');
});
server.listen(port, hostname, () => {
 console.log('Server running at http://${hostname}:${port}/');
});
```

示例中 index.js 文件使用的端口号为 6001，可根据实际需求自行修改。按 Esc 键，输入":wq"，保存文件并返回。

执行以下命令，运行 Node.js 项目：

```
node index.js
```

在本地浏览器中访问地址 http：127.0.0.1:6001，查看项目是否正常运行。

显示结果如图 1.14 所示，说明 Node.js 环境搭建成功。

图 1.14　Node.js 项目示例

1.6　Node.js 的开发调试工具 Inspect

本节介绍 Node.js 开发调试工具 Inspect。先来看一个例子。

新建名为 test.js 的文件，输入以下内容：

```
console.log('Hello Node.js');
```

那么如何执行文件呢？打开命令行工具，执行命令：

```
node test.js
```

可以看到输出结果：

```
$ node run.js
Hello Node.js
```

在实际开发中，我们会使用调试工具 Inspect。

（1）在命令行中执行如下命令，如图 1.15 所示。

```
node --inspect-brk test.js
```

第 1 章　Node.js 与 JavaScript

图 1.15　执行调试工具的命令

（2）使用 Chrome 浏览器，在地址栏输入如下命令，打开效果如图 1.16 所示。

```
chrome://inspect
```

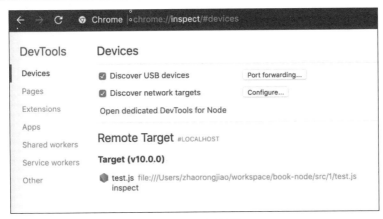

图 1.16　页面效果

（3）单击图 1.16 所示的 inspect，效果如图 1.17 所示。

图 1.17　inspect 效果

实际上，在 Node.js 执行代码时，Node.js 底层将代码进行封装。把封装的代码提取出来分析一下：

```
(function(exports, require, module, __filename, __dirname) {
    console.log(' Hello Node.js);
  }
);
```

Node.js 把代码封装在一个函数中并且添加了 5 个参数，前 3 个参数分别是：exports、require、module，实际上使用的是 CommonJS 模块的规范。

说　明
CommonJS 规范是 JavaScript 中的模块加载机制。在这个规范下，每个 .js 文件都是一个模块，它们内部各自使用的变量名和函数名都互不冲突。

第 2 章 NPM、REPL与Node.js工作原理

本章介绍 NPM、REPL，并重点介绍 Node.js 中同步/异步、单线程/多线程、并行/并发、事件循环等重要概念及其工作原理。这些核心概念是 Node.js 必不可少的组成部分，是使用 Node.js 的基石。

2.1 NPM 的使用

NPM 全称 Node Package Manager，即 Node 包管理器，它会随着 Node.js 自动安装，是 Node.js 默认的、以 JavaScript 编写的软件包管理系统。本节介绍 NPM 及其工作原理、属性说明和常用指令。

2.1.1 NPM 简介

NPM 是基于 Node.js 的包管理器，也是整个 Node.js 社区支持的、很流行的第三方模块的包管理器。NPM 适用于解决代码分享、重用和部署问题，包括第三方类库的导入和依赖关系的建立等问题，可以提升工程师的开发效率，其主要使用场景有：

- 允许用户从 NPM 服务器下载他人编写的第三方包到本地使用。
- 允许用户从 NPM 服务器下载并安装他人编写的命令行程序到本地使用。
- 允许用户将自己编写的包或命令行程序上传到 NPM 服务器供他人使用。

Node.js 已经集成了 NPM，因此 NPM 可随着 Node.js 一并安装完成，可以通过输入"npm -v"命令来测试是否成功安装：

```
$ npm -v
v6.9.0
```

如果你安装的是旧版本的 NPM，可以很容易地通过 NPM 命令来升级，命令如下：

```
npm install npm -g
```

2.1.2 NPM 的工作原理

对于包管理,首先要理解什么是包(Package)。包用于描述一个文件或一个目录。一个包的配置通常由以下部分构成:

- 一个文件夹包含一个 package.json 配置文件。
- 包含的 GZIP 压缩文件(含有 package.json 文件的文件夹)。
- 解析 GZIP 的 URL。
- 为注册表添加<name>@<version>的 URL 信息。

模块是将一个复杂的程序依据一定的规则(规范)封装成几个块(文件),并组合在一起。换句话说,模块是通过配置文件中的节点包含一个或多个包。通常是由包、配置文件以及相关模块构成完整的一个或多个业务功能单元。块的内部数据/实现是私有的,只是向外部暴露一些接口(方法)与外部其他模块通信。

NPM 允许在 package.json 文件中使用 scripts 字段定义脚本命令,例如:

```
{
// ...
"scripts":{
    "build":"node index.js"
}
}
```

这块代码是 package.json 文件的片段,其中 scripts 字段是一个对象,scripts 对象的每一个属性对应一段脚本。其中 build 命令对应的脚本是 node index.js。在命令行下运行 npm run 命令,就可以执行这段脚本:

```
$ npm run build
#等同于执行
$ node index.js
```

这些定义在 package.json 里面的脚本就称为 NPM 脚本。通过使用 NPM 脚本可以将项目中相关的脚本集中在一处。不同项目的脚本命令只要功能相同,就可以有同样的对外接口。同时,利用 NPM 可以提供很多辅助功能。

查看当前项目的所有 NPM 脚本命令,可以使用不带任何参数的 npm run 命令:

```
$ npm run
```

NPM 脚本的原理相对简单,每当执行 npm run 时,会自动新建一个 Shell,在该 Shell 中执行指定的脚本命令。因此,只要是 Shell(一般是 Bash)可以运行的命令,就可以写在 NPM 脚本里面。需要注意的是,npm run 新建的 Shell 会将当前目录的 node_modules/.bin 子目录加入 PATH 变量,命令执行结束后,再将 PATH 变量恢复。

也就是说,当前目录的 node_modules/.bin 子目录里面的所有脚本都可以直接用脚本名调

用,而不必加上路径。例如,当前项目的依赖里面有 Mocha,只需要直接写 mocha test 即可:

```
{
script: {
    "test":"mocha test"
}
}
```

而不需要写成:

```
{
script: {
    "test":"./node_modules/.bin/mocha test"
}
}
```

由于 NPM 脚本的唯一要求是可以在 Shell 中执行,因此它不一定是 Node 脚本,任何可执行文件都可以写在 script 中。NPM 脚本的退出码也遵守 Shell 脚本规则,如果退出码不是 0,NPM 就认为该脚本执行失败。

2.1.3 package.json 属性说明

NPM 在 package.json 文件中管理项目的依赖项以及项目的元数据。

node 执行 require 指令时,会根据 package.json 中的依赖项执行查找。每个项目的根目录下面,一般都有一个 package.json 文件,定义了这个项目所需要的各种模块,以及项目的配置信息(如项目名称、版本号、许可证等元数据)。npm install 命令会根据这个配置文件,自动下载所需的模块,也就是配置项目所需的运行和开发环境。例如:

```
{
  "name" : "editor",
  "version" : "1.0.0",
  "description" : "launch $EDITOR in your program",
  "main" : "index.js",
  "directories" : {
     "example" : "example",
     "test" : "test"
  },
  "dependencies" : {},
  "devDependencies" : {
     "tap" : "~0.4.4"
  },
  "scripts" : {
     "test" : "tap test/*.js"
  },
  "repository" : {
```

```
      "type" : "git",
      "url" : "git://github.com/substack/node-editor.git"
   },
   "homepage" : "https://github.com/substack/node-editor",
   "keywords" : [
      "text",
      "edit",
      "shell"
   ],
   "author" : {
      "name" : "James Halliday",
      "email" : "mail@substack.net",
      "url" : "http://substack.net"
   },
   "license" : "MIT",
   "engine" : { "node" : ">=0.6" }
}
```

package.json 文件就是一个 JSON 对象,该对象的每一个成员就是当前项目的一项基础设置信息,各成员说明如下:

- Name: 包名。
- Version: 包的版本号,语义版本号分为 X.Y.Z 三位,分别代表主版本号、次版本号和补丁版本号。
- Description: 包的描述。
- Homepage: 包的官网地址。
- Author: 包的作者姓名。
- Contributors: 包的其他贡献者姓名。
- Dependencies: 依赖包列表,指定了项目运行所依赖的模块。如果依赖包没有安装,npm 就会自动将依赖包安装在 node_module 目录下。
- devDependencies: 指定项目开发所需要的模块。
- repository: 包代码存放的地方的类型,可以是 Git 或 Svn,Git 可在 GitHub 上。
- main: main 字段指定了程序的主入口文件,require('moduleName') 就会加载这个文件。这个字段的默认值是模块根目录下面的 index.js。
- keywords: 关键字。
- scripts: 指定了运行脚本命令的 npm 命令行缩写,比如 start 指定了运行 npm run start 时所要执行的命令。
- bin: 用来指定各个内部命令对应的可执行文件的位置。
- config: 用于添加命令行的环境变量。

package.json 文件可以手工编写,也可以使用 npm init 命令自动生成:

```
npm init
```

这个命令采用交互方式要求用户回答一些问题，然后在当前目录生成一个基本的 package.json 文件。所有问题之中，只有项目名称（name）和项目版本（version）是必填的，其他都是选填的。

有了 package.json 文件，直接使用 npm install 命令就会在当前目录中安装所需要的模块。

```
$ npm install
```

如果一个模块不在 package.json 文件中，可以单独安装这个模块，并使用相应的参数将其写入 package.json 文件中：

```
$ npm install express --save
$ npm install express --save-dev
```

这两行代码表示单独安装 express 模块，--save 参数表示将该模块写入 dependencies 属性，--save-dev 表示将该模块写入 devDependencies 属性。

2.1.4　NPM 的常用命令

本小节介绍 NPM 使用过程中的常用命令，包括安装模块、卸载模块、更新模块、检查模块版本、查看安装的模块等。

（1）安装模块

使用 npm 命令安装 Node.js 模块语法格式如下：

```
npm install <Module Name>
```

例如，我们使用 npm 命令安装常用的 Node.js Web 框架模块 express：

```
$ npm install express -g        #全局安装 express
$ npm install express           #本地安装 express
```

npm 的包安装分为本地安装（local）和全局安装（global）两种。采用本地安装会将安装包放在 ./node_modules 下（运行 npm 命令时所在的目录），如果没有 node_modules 目录，就会在当前执行 npm 命令的目录下生成 node_modules 目录。同时，可以通过 require() 来引入本地安装的包。若采用全局安装，则会将安装包放在 /usr/local 下或者 Node.js 的安装目录下，可以直接在命令行里使用。

（2）卸载模块

```
npm uninstall [<@scope>/]<pkg>[@<version>]...
[-S|--save|-D|--save-dev|-O|--save-optional]
  aliases: remove, rm, r, un, unlink
```

例如卸载项目环境所依赖的模块：

```
npm uninstall gulp --save-dev
```

（3）更新模块

```
npm update express
```

（4）检查模块版本

```
npm outdated [[<@scope>/]<pkg> ...]
```

（5）查看安装的模块

查看安装信息：

```
npm list -g        #查看全局安装的模块
npm list           #查看本地安装的模块
```

也可以查看某个模块的信息：

```
npm list express
```

（6）搜索模块

```
npm search express
```

（7）查看某条命令的详细帮助

NPM 提供了很多命令，例如 install 和 publish，使用 npm help 可查看所有命令：

```
npm help
```

（8）查看包的安装路径

```
npm root [-g]
```

（9）管理模块的缓存

```
npm cache add <tarball file>
npm cache add <folder>
npm cache add <tarball url>
npm cache add <name>@<version>
npm cache ls [<path>]
npm cache clean [<path>]
```

常用命令是清除 NPM 本地缓存：

```
npm cache clean
```

（10）启动模块

```
npm start
```

（11）重新启动模块

```
npm restart
```

2.2 REPL 的使用

REPL（Read Eval Print Loop，交互式解释器）表示一个计算机环境，类似于 Windows 系统的终端或 UNIX/Linux Shell，可以在终端中输入命令，并接收系统的响应。

Node 自带了交互式解释器，可以执行以下任务：

- 读取：读取用户输入，解析输入的 JavaScript 数据结构并存储在内存中。
- 执行：执行输入的数据结构。
- 打印：输出结果。
- 循环：循环操作以上步骤直到用户两次按下 Ctrl+C 按钮退出。

通过 Node.js 的交互式解释器可以很好地调试 JavaScript 代码，可以输入以下命令来启动 Node 的终端：

```
$ node
>
```

在>后输入简单的表达式，并按回车键来计算结果，例如：

```
$ node
> 5 + 4
9
```

在 REPL 中，可以将数据存储在变量中，在需要的时候使用，例如：

```
$ node
> x = 10
10
> var y = 20
undefined
> x + y
30
> console.log("Hello World")
Hello World
undefined
```

【示例 2-1】Node REPL 支持输入多行表达式，例如执行一个 do-while 循环。

```
$ node
> var x = 0
undefined
> do {
... x++;
... console.log("x: " + x);
... } while ( x < 5 );
```

```
x: 1
x: 2
x: 3
x: 4
x: 5
undefined
>
```

其中 3 个点的符号是系统自动生成的，按回车键换行后即可。Node 会自动检测是否为连续的表达式。以上 Node REPL 输入与输出的交互式效果如图 2.1 所示。

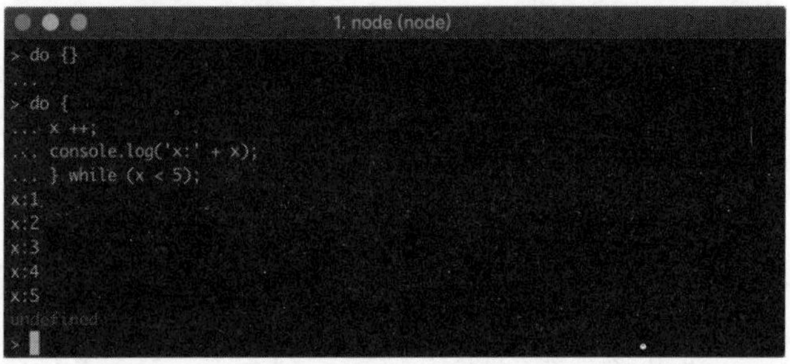

图 2.1　REPL 示例

REPL 的常用命令如下：

- Ctrl +C：退出当前终端。
- Ctrl +C 按下两次：退出 Node REPL。
- Ctrl +D：退出 Node REPL。
- 向上/向下键：查看输入的历史命令。
- Tab 键：列出当前命令。
- .help：列出使用命令。
- .break：退出多行表达式。
- .clear：退出多行表达式。
- .save filename：保存当前的 Node REPL 会话到指定文件。
- .load filename：载入当前 Node REPL 会话的文件内容。

2.3　Runtime 和 vm

本节介绍 Node.js 核心的两个概念：Runtime 和 vm。

2.3.1　Runtime

Runtime 又叫运行时，指将数据类型的确定由编译时推迟到了运行时。动态语言会将一些工作放在代码运行时才处理而并非编译时。也就是说,有很多类和成员变量在我们编译时是不知道的,而在运行时,我们所编写的代码会转换成完整的确定的代码运行。因此,编译器是不够的,我们还需要一个运行时系统(Runtime System)来处理编译后的代码。Runtime 基本是用 C 和汇编写的。

从概念上讲,JavaScript 引擎(engine)负责解析和 JIT 编译,例如把 JavaScript 中的语言编译成机器码。Runtime 提供内建的库,可以在程序运行时使用。所以可以在浏览器中使用 Window 对象或者 DOM API,这些存在于浏览器的 Runtime 中。Node.js Runtime 包含不同的库,如 Cluster 和 FileSystem API。两个 Runtime 都包含内置的数据类型和常用的工具,如 Console 对象。因此,Chrome 和 Node.js 共享相同的引擎(V8),但是它们具有不同的 Runtime。

因此,编写的代码具有运行时、动态特性。例如,在程序运行过程中,动态地创建类,动态地添加、修改该类的属性和方法;遍历一个类中所有的成员变量、属性以及所有方法;消息传递和转发。

Runtime 的典型事例:

- 给系统分类添加属性、方法。
- 方法交换。
- 获取对象的属性、私有属性。
- 字典转换模型。
- KVC、KVO。
- 归档(编码、解码)。
- NSClassFromString class<->字符串。
- block。
- 类的自我检测。

2.3.2 vm

vm 的概念比较广泛,通常可以认为是硬件和二进制文件的中间层。在 C++中,编译好的二进制文件可以直接被操作系统调用;而在 Java 中,编译好的字节码是交给虚拟机来运行的。这样的好处就是对开发者屏蔽了操作系统之间的差异,对于不同操作系统的处理交给了虚拟机来完成,从这个角度来看,vm 是对不同计算机系统的一种抽象。

vm 是 Node 的一个核心模块,vm 可以使用 V8 的 Virtual Machine Contexts 动态地编译和执行代码,而代码的执行上下文是与当前进程隔离的,但是这里的隔离并不是绝对安全的,不完全等同于浏览器的沙箱环境。vm 模块提供了一系列 API 用于在 V8 虚拟机环境中编译和运行代码。JavaScript 代码可以被编译并立即运行,或编译、保存,然后运行。

【示例 2-2】

```
const util = require('util');
const vm = require('vm');
```

```javascript
// 1. 创建一个 vm.Script 实例，编译要执行的代码
const script = new vm.Script('globalVar += 1; anotherGlobalVar = 1; ');
// 2. 用于绑定到 context 的对象
const sandbox = {globalVar: 1};
// 3. 创建一个 context，并且把 sandbox 对象绑定到这个环境，作为全局对象
const contextifiedSandbox = vm.createContext(sandbox);
// 4. 运行上面编译的代码，context 是 contextifiedSandbox
const result = script.runInContext(contextifiedSandbox);

console.log('sandbox === contextifiedSandbox ? ${sandbox ===www.bsck.org contextifiedSandbox}');
// sandbox === contextifiedSandbox ? true
console.log('sandbox: ${util.inspect(sandbox)}');
// sandbox: { globalVar: 2, anotherGlobalVar: 1 }
console.log('result: ${util.inspect(result)}');
// result: 1
```

vm.Script 是一个类，用于创建代码实例，后面可以多次运行。vm.createContext(sandbox) 用于 contextify 一个对象，根据 ECMAScript 2015 语言规范，代码的执行需要一个 execution context。这里的 contextify 就是把传进去的对象与 V8 的一个新的 context 进行关联。这里所说的关联是指 contextified 对象的属性将会成为 context 的全局属性，同时，在 context 下运行代码时产生的全局属性也会成为这个 contextified 对象的属性。

script.runInContext(contextifiedSandbox)就是使代码在 contextifiedSandbox 这个 context 中运行。从上面的输出可以看到，代码运行后，contextifiedSandbox 中的属性的值已经被改变了，运行结果显示的是最后一个表达式的值。

除了上面几个接口之外，vm 模块还有一些更便捷的接口，例如 vm.runInContext(code, contextifiedSandbox[,www.90168.org options])、vm.runInNewContext(code[,sandbox][,options])等。

2.4 回调函数

Node.js 使用了大量的回调函数，其异步编程的直接体现就是回调函数。Node.js 在完成任务后会调用回调函数，Node.js 几乎每一个 API 都是支持回调函数的。

在 Node 应用程序中，执行异步操作的函数将回调函数作为最后一个参数，回调函数接收错误对象作为第一个参数：

```javascript
function foo1(name, age, callback) {
 // do something
}
```

```
function foo2(value, callback1, callback2) {
  // do something
}
```

其中，callback 就是回调函数。例如，可以一边读取文件，一边执行其他命令，在文件读取完成后，将文件内容作为回调函数的参数返回。这样在执行代码时就没有阻塞或等待文件 I/O 操作，也就大大提高了 Node.js 的性能，可以处理大量的并发请求。

【示例 2-3】让我们来看个具体的示例，创建一个 test.txt 文件，该文件内容如下：

```
Hello World!
```

创建 main.js 文件，代码如下：

```
var fs = require("fs");

fs.readFile(test.txt', function (err, data) {
  if (err){
    console.log(err.stack);
    return;
  }
  console.log(data.toString());
});
console.log("Done !");
```

以上程序中，fs.readFile() 是用于读取文件的异步函数。如果在读取文件的过程中发生错误，错误 err 对象就会输出错误信息。如果没有发生错误，readFile 跳过 err 对象的输出，文件内容就通过回调函数输出。执行以上代码，执行结果如下：

```
Done!
Hello World!
```

接下来我们删除 test.txt 文件，执行结果如下：

```
Done!
Error: ENOENT, open 'input.txt'
```

因为文件 test.txt 不存在，所以输出了错误信息。从上述示例代码可以看出，fs.readFile()方法的最后一个参数就是回调函数，而回调函数里面第一个参数就是回调函数接收的错误参数。

2.5 同步/异步和阻塞/非阻塞

经常可以看到有人将同步和阻塞等同，异步和非阻塞等同。事实上，这两对概念有一定的区别，不能混淆。

2.5.1 同步和异步

首先，理解同步和异步的概念。同步就是指一个进程在执行某个请求的时候，若该请求需要一段时间才能返回信息，则这个进程将会一直等待下去，直到收到返回信息才继续执行下去；异步是指进程不需要一直等下去，而是继续执行下面的操作，不管其他进程的状态。当有消息返回时，系统会通知进程进行处理，这样可以提高执行的效率。

由于 JavaScript 是单线程模型，执行 IO 操作时，JavaScript 代码无须等待，而是传入回调函数后，继续执行后续的 JavaScript 代码。比如 jQuery 提供的 getJSON()操作：

```
$.getJSON('http://example.com/ajax', function (data) {
    console.log('IO 结果返回后执行...');
});
console.log('不等待 IO 结果直接执行后续代码...');
```

而同步的 IO 操作则需要等待函数返回：

```
// 根据网络耗时，函数将执行几十毫秒到几秒不等
var data = getJSONSync('http://example.com/ajax');
```

同步操作的好处是代码简单，缺点是程序将等待 IO 操作，在等待时间内，无法响应其他任何事件，而异步读取则不需要等待 IO 操作。

还可以使用 async/await 来处理异步。async 函数返回一个 Promise 对象，可以使用 then 方法添加回调函数。当函数执行时，遇到 await 就会等待其异步操作完成，然后执行函数体后面的语句。async 放在函数前，表示函数里有异步操作，例如：

```
async function foo(){

} //函数声明
```

或：

```
const foo = async function foo(){

} // 函数表达式
```

async 函数的返回值为 Promise 对象：

```
async function f(){
  return 'hello world';
}

f().then(v => console.log(v));
// "hello world"

async function f(){
  throw new Error('报错了'); // 返回的 Promise 对象为 reject 状态
```

```
}
f().then(
  v => console.log(v),
  e => console.log(e)
);
// Error: 报错了
// 用 catch 接受错误信息
async function f(){
  throw new Error('报错了');
}
f()
.then(v => console.log(v))
.catch(e => console.log(e))
// Error: 报错了
```

await 表示紧跟在后面的表达式需要等待结果。一般为 Promise 对象，如果不是，就会被转成一个立即 resolve 的 Promise 对象。

【示例 2-4】防止异步函数出错，需要把 await 命令放在 try…catch 代码块中：

```
function test2(){
  setTimeout(()=>{
    console.log(1000);
  },1000)
}

async function test(){
  for(let i = 0; i < 5; i++){
    try {
      await test2(); // 异步函数
      break

    } catch (e) {
      console.log('a', e);
    }
  }

}
test();
```

同步问题多发生在多线程环境的数据共享问题中,即当多个线程需要访问同一个资源时,它们需要以某种顺序来确保该资源在某一特定时刻只能被一个线程访问,如果使用异步,程序的运行结果将不可预料。因此,在这种情况下,就必须对数据进行同步,即限制只能有一个进程访问资源,其他线程必须等待。

2.5.2 阻塞和非阻塞

让我们先了解阻塞（Blocking）和非阻塞（Non-Blocking）的区别。阻塞调用是指调用结果返回之前，当前线程会被挂起。函数只有在得到结果之后才会返回。非阻塞和阻塞的概念相对应，指在不能立刻得到结果之前，该函数不会阻塞当前线程，而会立刻返回。

【示例 2-5】看一个阻塞代码的示例。

首先创建一个文件 input.txt，内容如下：

```
Hello Sync!
```

创建 main1.js 文件，代码如下：

```
var fs = require("fs");

var data = fs.readFileSync('input.txt');

console.log(data.toString());
console.log("Done!");
```

以上代码执行结果如下：

```
node main.js
Hello Sync!
Done!
```

【示例 2-6】再看一个非阻塞代码示例。

创建一个文件 test1.txt，内容如下：

```
Hello Async!
```

创建 main2.js 文件，代码如下：

```
var fs = require("fs");

fs.readFile(test1.txt', function (err, data) {
    if (err) return console.error(err);
    console.log(data.toString());
});

console.log("Done!");
```

以上代码执行结果如下：

```
node main.js
Done!
Hello Async!
```

通过以上两个示例我们了解了阻塞与非阻塞调用的不同。第一个示例在文件读取完后才执行完程序。第二个示例我们不需要等待文件读取完，就可以在读取文件时同时执行接下来的代码，大大提高了程序的性能。

因此，阻塞是按顺序执行的，而非阻塞是不需要按顺序的，所以如果需要处理回调函数的参数，就需要写在回调函数内。

2.5.3 同步/异步和阻塞/非阻塞

同步和异步是相对于操作结果来说的，是否会等待结果返回。阻塞和非阻塞是相对于线程是否被阻塞来说的。

有人也许会把阻塞调用和同步调用等同起来，实际上它们是不同的。其实，这两者存在本质的区别，它们的修饰对象是不同的。阻塞和非阻塞是指进程访问的数据如果尚未就绪，进程是否需要等待，简单来说，这相当于函数内部的实现区别，也就是未就绪时是直接返回还是等待就绪。而同步和异步是指访问数据的机制，同步一般指主动请求并等待IO操作完毕的方式，当数据就绪后，在读写的时候必须阻塞，异步则指主动请求数据后便可以继续处理其他任务，随后等待IO操作完毕的通知，这可以使进程在数据读写时也不阻塞。

对于同步调用来说，很多时候当前线程还是激活的，只是从逻辑上当前函数没有返回而已。当工作在阻塞模式的时候，如果在没有数据的情况下调用该函数，当前线程就会被挂起，直到有数据为止。

2.6 单线程和多线程

当一个程序开始运行时，它就是一个进程，进程包括运行中的程序和程序所使用到的内存和系统资源。而一个进程又是由多个线程所组成的。线程是程序中的一个执行流，每个线程都有自己的专有寄存器（栈指针、程序计数器等），但代码区是共享的，即不同的线程可以执行同样的函数。多线程是指程序中包含多个执行流，即在一个程序中可以同时运行多个不同的线程来执行不同的任务。也就是说，允许单个程序创建多个并行执行的线程来完成各自的任务。多线程可以提高 CPU 的利用率。在多线程程序中，一个线程必须等待的时候，CPU 可以运行其他的线程而不是等待，这样就大大提高了程序的效率。

Node.js 是单线程的指的是 JavaScript 的执行是单线程的，但 JavaScript 的宿主环境，无论是 Node 还是浏览器都是多线程的。Node.js 基于 V8 引擎，所以它本身并不支持多线程，但是为了充分利用 Server 的 Multi-Core，必须使用多进程的方式。

Node.js 与进程相关的模块有 process、child_process、cluster，其中 cluster 用于方便地创建共享端口的多进程模式。

2.7 并行和并发

1. 什么是并发

并发（Concurrency）指应用能够交替执行不同的任务，类似于多线程，多线程并非是同时执行多个任务，而是以几乎不可察觉的速度在多个任务之间不断切换，以达到一种"同时执行"的效果，而事实上并不是同时执行，只是切换速度快到无法察觉。

2. 什么是并行

并行（Parallellism）指应用能够同时执行不同的任务，不同的任务是可以同时执行的。

并发与并行两者很重要的区别是：并发是交替执行的，而并行是同时执行的。通俗地理解，并发是不同的代码块交替执行，并行是不同的代码块同时执行。如果某个系统支持两个或者多个动作（Action）同时存在，那么这个系统就是一个并发系统。如果某个系统支持两个或者多个动作同时执行，那么这个系统就是一个并行系统。

并行是指两个或者多个事件在同一时刻发生。而并发是指两个或多个事件在同一时间间隔发生。并行是在不同实体上的多个事件，并发是在同一实体上的多个事件。并行是在一台处理器上"同时"处理多个任务，而并发是在多台处理器上同时处理多个任务。所以并发编程的目标是充分地利用处理器的每一个核，以达到最高的处理性能。并发强调的是一起出发，并行强调的是一起执行。并发的反义是顺序，并行的反义是串行。并发并行并不是互斥概念，只不过并发强调任务的抽象调度，并行强调任务的实际执行。

单线程解决高并发问题的方法就是采用非阻塞、异步编程的思想。当遇到非常耗时的 IO 操作时，采用非阻塞的方式继续执行后面的代码，并且进入事件循环，当 IO 操作完成时，程序会被通知 IO 操作已经完成。这个主要运用 JavaScript 的回调函数来实现。

多线程虽然也能解决高并发，但是以建立多个线程来实现，其缺点是当遇到耗时的 IO 操作时，当前线程会被阻塞，并且把 CPU 的控制权交给其他线程，这样带来的问题就是要非常频繁地进行线程的上下文切换。

并发系统与并行系统这两个定义之间的关键差异在于"存在"这个词。

Node.js 是单进程单线程应用程序，但是因为 V8 引擎提供的异步执行回调接口，通过这些接口可以处理大量的并发，所以性能非常高。Node.js 是单线程且支持高并发的脚本语言，IO 密集型处理是 Node.js 的强项，因为 Node.js 的 IO 请求都是异步的。对于异步，发出操作指令，然后就可以去做别的事情了（主线程无须等待），所有操作完成后执行回调。

【示例 2-7】

```
let a = 1; // step1：定义变量

// step2：发出指令，然后把回调函数加入事件队列（回调函数并没有执行）
```

```
setTimeout(() => {
    console.log(a);
}, 0)

a = 2; // step3：赋值，回调函数没有执行

// step4：发出指令，然后把回调函数加入异步队列（回调函数并没有执行）
setTimeout(() => {
    console.log(a);
}, 0)

a = 3; // step5：赋值，回调函数没有执行

// 当所有代码执行完毕，cpu 空闲下来了，就会开始遍历执行事件队列里面的回调函数
// 最后输出：3 3
```

异步 I/O 的 Node.js 为什么可以支持高并发？因为 IO 操作由 Node.js 的工作线程执行，Node.js 底层的 libuv 是通过多线程的线程池来并行 IO 操作的，主线程不需要等待结果返回，发出指令后就去执行其他事务。

3. 如何处理高并发和非阻塞请求

针对每个并发请求，服务端给请求注册一个激发事件（I/O），并给一个回调函数（这个过程没有阻塞新的连接请求）。按顺序执行事件处理（I/O），处理完成后执行回调函数，接着执行下一个事件处理（I/O）。事件处理（I/O）的原理是什么呢？事件处理（异步 I/O 处理）是由 Node 工作线程去执行的（Node.js 底层的 libuv 是通过多线程的线程池来并行 I/O 操作的），且主线程不需要等待返回，只要发出指令后就可以执行其他事件，所有操作完成后执行回调。

2.8 事件循环

Node.js 基本上所有的事件机制都是使用设计模式中的观察者模式实现的。Node.js 单线程类似于进入一个 while(true) 的事件循环，直到没有事件观察者退出，每个异步事件都生成一个事件观察者，如果有事件发生，就调用该回调函数。

Node.js 使用事件驱动模型，当 Web Server 接收到请求时，就把它关闭，然后进行处理，接着去服务下一个 Web 请求。当这个请求完成后，它被放回处理队列，当到达队列开头后，这个结果被返回给用户。

这个模型非常高效，可扩展性非常强，因为 Web Server 一直接受请求而不等待任何读写操作。这也被称为非阻塞式 IO 或者事件驱动 IO。

在事件驱动模型中会生成一个主循环来监听事件，当检测到事件时触发回调函数。

整个事件驱动的流程如图 2.2 所示。事件循环模型非常简洁，类似于观察者模式，事件相当于一个主题（Subject），而所有注册到这个事件上的处理函数相当于观察者（Observer）。

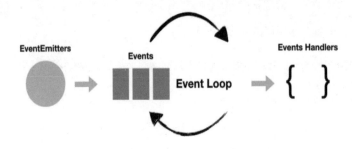

图 2.2 事件驱动的流程

Node.js 有多个内置的事件，我们可以通过引入 events 模块，并通过实例化 EventEmitter 类来绑定和监听事件，示例如下：

```
// 引入 events 模块
var events = require('events');
// 创建 eventEmitter 对象
var eventEmitter = new events.EventEmitter();
```

以下程序绑定事件处理程序：

```
// 绑定事件及事件的处理程序
eventEmitter.on('eventName', eventHandler);
```

我们可以通过程序触发事件：

```
// 触发事件
eventEmitter.emit('eventName');
```

【示例 2-8】完整的示例代码如下，创建 main3.js 文件：

```
// 引入 events 模块
var events = require('events');
// 创建 eventEmitter 对象
var eventEmitter = new events.EventEmitter();

// 创建事件处理程序
var connectHandler = function connected() {
   console.log('连接成功。');

   // 触发 data_received 事件
   eventEmitter.emit('data_received');
}

// 绑定 connection 事件处理程序
eventEmitter.on('connection', connectHandler);
```

```
// 使用匿名函数绑定 data_received 事件
eventEmitter.on('data_received', function(){
   console.log('数据接收成功。');
});

// 触发 connection 事件
eventEmitter.emit('connection');

console.log("程序执行完毕。");
```

接下来执行以上代码,结果如下:

```
$ node main3.js
连接成功。
数据接收成功。
程序执行完毕。
```

第 3 章
Node.js的语法

Node.js 的基础是 JavaScript 脚本语言，Node.js 的语法还是 JavaScript 语法，只不过它封装了一些类库，提供给开发者，其他由运行环境去处理。JavaScript 是一门脚本语言，脚本语言都需要一个解析器才能运行。对于写在 HTML 页面里的 JS，对应的解析器就是浏览器。而对于需要独立运行的 JavaScript，Node.js 就是一个解析器。

3.1 ECMAScript 6 标准

本章开始前，读者务必先了解 ECMAScript 6 标准。

ECMAScript 6（以下简称 ES6）是 JavaScript 语言的标准，在 2015 年 6 月正式发布。ES6 的目标是使得 JavaScript 语言可以用来编写复杂的大型应用程序，成为企业级开发语言。

ECMAScript 是浏览器脚本语言的标准，从一开始就是针对 JavaScript 语言制定的。ES6 的第一个版本于 2015 年 6 月发布，正式名称为《ECMAScript 2015 标准》（简称 ES 2015）。ES6 也是一个泛指，指的是 5.1 版以后的 JavaScript 的下一代标准，涵盖了 ES 2015、ES 2016、ES 2017 等，泛指"下一代 JavaScript 语言"。

3.2 数组常用方法及 ES6 中的数组方法

本节介绍数组方法，包括常用的数组方法以及 ES6 中新增的数组方法。

（1）arr.push() 在数组末尾添加元素，返回值为添加完后的数组的长度，例如：

```
let arr = [1,2,3,4,5]
console.log(arr.push(5))      // 6
console.log(arr)              // [1,2,3,4,5,5]
```

（2）arr.pop() 从数组末尾删除元素，只能是一个，返回值是删除的元素，例如：

```
let arr = [1,2,3,4,5]
```

```
console.log(arr.pop())        // 5
console.log(arr)              //[1,2,3,4]
```

(3) arr.shift() 从数组起始位置删除元素，只能删除一个，返回值是删除的元素，例如：

```
let arr = [1,2,3,4,5]
console.log(arr.shift())      // 1
console.log(arr)              // [2,3,4,5]
```

(4) arr.unshift() 从数组起始位置添加元素，返回值是添加完后的数组的长度，例如：

```
let arr = [1,2,3,4,5]
console.log(arr.unshift(2))   // 6
console.log(arr)              //[2,1,2,3,4,5]
```

(5) arr.splice(i,n) 删除从 i（索引值）开始之后的元素，返回值是被删除的元素，参数 i 是索引值，n 为个数，例如：

```
let arr = [1,2,3,4,5]
console.log(arr.splice(2,2))  //[3,4]
console.log(arr)              // [1,2,5]
```

(6) arr.concat()连接两个数组，返回值为连接后的新数组，例如：

```
let arr = [1,2,3,4,5]
console.log(arr.concat([1,2])) // [1,2,3,4,5,1,2]
console.log(arr)              // [1,2,3,4,5]
```

(7) str.split()将字符串转化为数组，例如：

```
let str = '123456'
console.log(str.split(''))    //["1", "2", "3", "4", "5", "6"]
```

(8) arr.sort()将数组进行排序，返回值是排序之后的数组，默认是按照最左边的数字进行排序的，不是按照数字大小排序的，例如：

```
let arr = [2,10,6,1,4,22,3]
console.log(arr.sort())       //[1, 10, 2, 22, 3, 4, 6]
let arr1 = arr.sort((a, b) =>a - b)
console.log(arr1)             //[1, 2, 3, 4, 6, 10, 22]
let arr2 = arr.sort((a, b) =>b-a)
console.log(arr2)             //[22, 10, 6, 4, 3, 2, 1]
```

(9) arr.reverse()将数组反转，返回值是反转后的数组，例如：

```
let arr = [1,2,3,4,5]
console.log(arr.reverse())    //[5,4,3,2,1]
console.log(arr)              //[5,4,3,2,1]
```

（10）arr.slice(start,end) 删除索引值 start 到索引值 end 的数组，不包含 end 索引的值，返回值是删除之后的数组，例如：

```
let arr = [1,2,3,4,5]
console.log(arr.slice(1,3))      // [2,3]
console.log(arr)                 // [1,2,3,4,5]
```

（11）arr.forEach(callback)遍历数组，无返回值。

callback 的参数如下：

- value：当前索引的值。
- index：索引。
- array：原数组。

【示例 3-1】

```
let arr = [1,2,3,4,5]
arr.forEach( (value,index,array)=>{
    console.log('value:${value}    index:${index}    array:${array}')
})
// value:1    index:0    array:1,2,3,4,5
// value:2    index:1    array:1,2,3,4,5
// value:3    index:2    array:1,2,3,4,5
// value:4    index:3    array:1,2,3,4,5
// value:5    index:4    array:1,2,3,4,5

let arr = [1,2,3,4,5]
arr.forEach( (value,index,array)=>{
    value = value * 2
    console.log('value:${value}    index:${index}    array:${array}')
})
console.log(arr)
// value:2     index:0    array:1,2,3,4,5
// value:4     index:1    array:1,2,3,4,5
// value:6     index:2    array:1,2,3,4,5
// value:8     index:3    array:1,2,3,4,5
// value:10    index:4    array:1,2,3,4,5
// [1, 2, 3, 4, 5]
```

（12）arr.map(callback) 映射数组（遍历数组），由 return 返回一个新数组，例如：

callback 的参数如下：

- value：当前索引的值。
- index：索引。
- array：原数组。

【示例 3-2】
```
let arr = [1,2,3,4,5]
arr.map( (value,index,array)=>{
     value = value * 2
     console.log('value:${value}    index:${index}    array:${array}')
})
console.log(arr)
```

其中，注意 arr.forEach()和 arr.map()的区别：

- arr.forEach()与 for 循环相同，可替代 for 循环使用。arr.map()是修改数组中的数据，并返回新的数据。
- arr.forEach() 没有 return 值，而 arr.map() 有 return 值。

（13）arr.filter(callback) 过滤数组，返回满足要求的新数组，例如：

```
let arr = [1,2,3,4,5]
let arr1 = arr.filter( (i, v) => i < 3)
console.log(arr1)         // [1, 2]
```

（14）arr.every(callback) 依据条件判断数组的元素是否全部满足，若满足，则返回 ture，例如：

```
let arr = [1,2,3,4,5]
let arr1 = arr.every( (i, v) => i < 3)
console.log(arr1)      // false
let arr2 = arr.every( (i, v) => i < 10)
console.log(arr2)      // true
```

（15）arr.some() 依据条件判断数组的元素是否有一个满足，若有一个满足，则返回 ture，例如：

```
let arr = [1,2,3,4,5]
let arr1 = arr.some( (i, v) => i < 3)
console.log(arr1)           // true
let arr2 = arr.some( (i, v) => i > 10)
console.log(arr2)           // false
```

（16）arr.reduce(callback, initialValue) 迭代数组的所有项，是一个累加器，数组中的每个值（从左到右）合并，最终计算为一个值。

参数说明如下：

- callback:previousValue（必选）：上一次调用回调返回的值，或者是提供的初始值（initialValue）。
- currentValue（必选）：数组中当前被处理的数组项。
- index（可选）：当前数组项在数组中的索引值。

- array（可选）：原数组。
- initialValue（可选）：初始值。

实现方法：回调函数第一次执行时，preValue 和 curValue 可以是一个值，如果 initialValue 在调用 reduce()时被提供，那么第一个 preValue 等于 initialValue ，并且 curValue 等于数组中的第一个值；如果 initialValue 未被提供，那么 preValue 等于数组中的第一个值。

【示例 3-3】

```
let arr = [0,1,2,3,4]
let arr1 = arr.reduce((preValue, curValue) =>
    preValue + curValue
)
console.log(arr1)    // 10


let arr2 = arr.reduce((preValue,curValue)=>preValue + curValue,5)
console.log(arr2)    // 15
```

（17）Array.from() 将伪数组变成数组，就是只要有 length 就可以转成数组。这是 ES6 中新增的标准方法，例如：

```
let str = '12345'
console.log(Array.from(str))    // ["1", "2", "3", "4", "5"]
let obj = {0:'a',1:'b',length:2}
console.log(Array.from(obj))    // ["a", "b"]
```

（18）Array.of() 将一组值转换成数组，类似于声明数组。这是 ES6 中新增的标准方法，例如：

```
let str = '11'
console.log(Array.of(str))        // ['11']
```

等价于：

```
console.log(new Array('11'))  // ['11]
```

注意 new Array()有个缺点，就是参数问题引起的重载：

```
console.log(new Array(2))    //[empty × 2]是一个空数组
console.log(Array.of(2))     // [2]
```

（19） arr.copyWithin() 在当前数组内部，将指定位置的数组复制到其他位置，会覆盖原数组项，返回当前数组。参数说明如下：

- target（必选）：索引从该位置开始替换数组项。
- start（可选）：索引从该位置开始读取数组项，默认为 0。如果为负值，就从右往左读。

- end（可选）：索引到该位置停止读取的数组项，默认是 Array.length。如果是负值，就表示倒数。

【示例 3-4】

```
let arr = [1,2,3,4,5,6,7]
let arr1 = arr.copyWithin(1)
console.log(arr1)    // [1, 1, 2, 3, 4, 5, 6]
let arr2 = arr.copyWithin(1,2)
console.log(arr2)    // [1, 3, 4, 5, 6, 7, 7]
let arr3 = arr.copyWithin(1,2,4)
console.log(arr3)    // [1, 3, 4, 4, 5, 6, 7]
```

（20）arr.find(callback) 找到第一个符合条件的数组成员，例如：

```
let arr = [1,2,3,4,5,2,4]
let arr1 = arr.find((value, index, array) =>value > 2)
console.log(arr1)    // 3
```

（21）arr.findIndex(callback) 找到第一个符合条件的数组成员的索引值，例如：

```
let arr = [1,2,3,4,5]
let arr1 = arr.findIndex((value, index, array) => value > 3)
console.log(arr1)    // 3
```

（22）arr.fill(target, start, end) 使用给定的值，填充一个数组，填充完后会改变原数组。参数说明如下：

- target: 待填充的元素。
- start: 开始填充的位置索引。
- end: 终止填充的位置索引（不包括该位置）。

【示例 3-5】

```
let arr = [1,2,3,4,5]
let arr1 = arr.fill(5)
console.log(arr1)       // [5, 5, 5, 5, 5]
console.log(arr)        // [5, 5, 5, 5, 5]
let arr2 = arr.fill(5,2)
console.log(arr2)
let arr3 = arr.fill(5,1,3)
console.log(arr3)
```

（23）arr.includes() 判断数中是否包含给定的值：

```
let arr = [1,2,3,4,5]
let arr1 = arr.includes(2)
console.log(arr1)       // ture
let arr2 = arr.includes(9)
```

```
console.log(arr2)          // false
let arr3 = [1,2,3,NaN].includes(NaN)
console.log(arr3)          // true
```

includes()与indexOf()的区别：

- indexOf()返回的是数值，而includes()返回的是布尔值。
- indexOf()不能判断NaN，返回为-1，includes()则可以判断。

（24）arr.keys()遍历数组的键名：

```
let arr = [1,2,3,4]
let arr2 = arr.keys()
for (let key of arr2) {
    console.log(key);    // 0,1,2,3
}
```

（25）arr.values() 遍历数组键值：

```
let arr = [1,2,3,4]
let arr1 = arr.values()
for (let val of arr1) {
    console.log(val);    // 1,2,3,4
}
```

（26）arr.entries() 遍历数组的键名和键值：

```
let arr = [1,2,3,4]
let arr1 = arr.entries()
for (let e of arr1) {
    console.log(e);     // [0,1] [1,2] [2,3] [3,4]
}
```

3.3 函数

ES6中的函数增加了很多新特性，使得在JavaScript中编程不容易出错，且比以往更加强大。ES6的函数主要带来了以下扩展：参数的默认值、rest参数、解构参数、扩展（Spread）运算符、name属性、箭头函数等。

3.3.1 参数的默认值

在ES6之前，不能直接为函数的参数指定默认值，只能采用变通的方法，例如：

```
function log(x, y) {
  y = y || 'World';
```

```
  console.log(x, y);
}

log('Hello')              // Hello World
log('Hello', 'China')     // Hello China
log('Hello', '')          // Hello World
```

上面的代码检查函数 log 的参数 y 有没有赋值，如果没有，就指定默认值为 World。这种写法的缺点在于，如果参数 y 赋值了，但是对应的布尔值为 false，该赋值就不起作用。就像上面代码的最后一行，参数 y 等于空字符，结果被改为默认值。

为了避免这个问题，通常需要先判断一下参数 y 是否被赋值，如果没有，再等于默认值。

```
if (typeof y === 'undefined') {
  y = 'World';
}
```

ES6 允许为函数的参数设置默认值，即直接写在参数定义的后面，例如：

```
function log(x, y = 'World') {
  console.log(x, y);
}

log('Hello')              // Hello World
log('Hello', 'China')     // Hello China
log('Hello', '')          // Hello
```

可以看到，ES6 的写法比 ES5 简洁许多，而且非常自然，例如：

```
function Point(x = 0, y = 0) {
  this.x = x;
  this.y = y;
}

var p = new Point();
p // { x: 0, y: 0 }
```

除了简洁，ES6 的写法还有两个好处：首先，阅读代码的人可以立刻意识到哪些参数是可以省略的，不用查看函数体或文档；其次，有利于将来代码的优化，即使未来的版本在对外接口中彻底拿掉这个参数，也不会导致以前的代码无法运行。

参数变量是默认声明的，所以不能用 let 或 const 再次声明。

```
function foo(x = 5) {
  let x = 1; // error
  const x = 2; // error
}
```

上面的代码中，参数变量 x 是默认声明的，在函数体中，不能用 let 或 const 再次声明，否则会报错。

使用函数默认参数时,不允许有同名参数,否则会报错:

```
// 不报错
function fn(name,name){
 console.log(name);
}

// 报错
//SyntaxError: Duplicate parameter name not allowed in this context
function fn(name,name,age=17){
 console.log(name+","+age);
}
```

只有在未传递参数或者参数为 undefined 时,才会使用默认参数,null 值被认为是有效的值传递:

```
function fn(name,age=17){
    console.log(name+","+age);
}
fn("Amy",null); // Amy,null
```

参数默认值可以与解构赋值的默认值结合起来使用,例如:

```
function foo({x, y = 5}) {
  console.log(x, y);
}

foo({}) // undefined, 5
foo({x: 1}) // 1, 5
foo({x: 1, y: 2}) // 1, 2
foo() // TypeError: Cannot read property 'x' of undefined
```

上面的代码使用了对象的解构赋值默认值,而没有使用函数参数的默认值。只有当函数 foo 的参数是一个对象时,变量 x 和 y 才会通过解构赋值而生成。如果函数 foo 调用时参数不是对象,变量 x 和 y 就不会生成,从而报错。如果参数对象没有 y 属性,y 的默认值 5 才会生效。

再来看看下面两种写法有什么差别?

```
// 写法一
function m1({x = 0, y = 0} = {}) {
  return [x, y];
}

// 写法二
function m2({x, y} = { x: 0, y: 0 }) {
```

```
    return [x, y];
}
```

上面两种写法都对函数的参数设定了默认值,区别是写法一函数参数的默认值是空对象,但是设置了对象解构赋值的默认值;写法二函数参数的默认值是一个有具体属性的对象,但是没有设置对象解构赋值的默认值。在使用中的差异:

```
// 函数没有参数的情况
m1() // [0, 0]
m2() // [0, 0]

// x 和 y 都有值的情况
m1({x: 3, y: 8}) // [3, 8]
m2({x: 3, y: 8}) // [3, 8]

// x 有值、y 无值的情况
m1({x: 3}) // [3, 0]
m2({x: 3}) // [3, undefined]

// x 和 y 都无值的情况
m1({}) // [0, 0];
m2({}) // [undefined, undefined]

m1({z: 3}) // [0, 0]
m2({z: 3}) // [undefined, undefined]
```

通常情况下,定义了默认值的参数应该是函数的尾参数,因为这样比较容易看出来到底省略了哪些参数。如果非尾部的参数设置默认值,实际上这个参数是没法省略的。

```
// 例一
function f(x = 1, y) {
  return [x, y];
}

f()         // [1, undefined]
f(2)        // [2, undefined])
f(, 1)      // 报错
f(undefined, 1) // [1, 1]

// 例二
function f(x, y = 5, z) {
  return [x, y, z];
}

f()         // [undefined, 5, undefined]
f(1)        // [1, 5, undefined]
```

```
f(1, ,2) // 报错
f(1, undefined, 2) // [1, 5, 2]
```

上面的代码中,有默认值的参数都不是尾参数。这时,无法只省略该参数,而不省略它后面的参数,除非显式地输入 undefined。

函数参数默认值存在暂时性死区,在函数参数默认值表达式中,还未初始化赋值的参数值无法作为其他参数的默认值。

```
function f(x,y=x){
    console.log(x,y);
}
f(1); // 1 1

function f(x=y){
    console.log(x);
}
f(); // ReferenceError: y is not defined
```

不定参数用来表示不确定参数个数,形如...变量名,由...加上一个具名参数标识符组成。具名参数只能放在参数组的最后,并且有且只有一个不定参数。

```
function f(...values){
    console.log(values.length);
}
f(1,2);       //2
f(1,2,3,4);   //4
```

3.3.2 rest 参数

ES6 引入了 rest 参数(形式为...变量名),用于获取函数的多余参数,这样就不需要使用 arguments 对象了。

【示例 3-6】rest 参数搭配的变量是一个数组,该变量将多余的参数放入数组中:

```
function add(...values) {
  let sum = 0;

  for (var val of values) {
    sum += val;
  }

  return sum;
}

add(2, 5, 3) // 10
```

上面代码的 add 函数是一个求和函数,利用 rest 参数可以向该函数传入任意数目的参数。

【示例 3-7】使用 rest 参数代替 arguments 变量：

```
// arguments 变量的写法
function sortNumbers() {
  return Array.prototype.slice.call(arguments).sort();
}

// rest 参数的写法
const sortNumbers = (...numbers) => numbers.sort();
```

上面代码的两种写法，比较后可以发现，rest 参数的写法更简洁。arguments 对象不是数组，而是一个类似数组的对象。所以为了使用数组的方法，必须使用 Array.prototype.slice.call 先将其转为数组。rest 参数就是一个真正的数组，数组特有的方法都可以使用。

【示例 3-8】再比如，使用 rest 参数改写数组 push 方法：

```
function push(array, ...items) {
  items.forEach(function(item) {
    array.push(item);
    console.log(item);
  });
}

var a = [];
push(a, 1, 2, 3)
```

注意，rest 参数之后不能再有其他参数，也就是说 rest 参数只能是最后一个参数，否则会报错：

```
// 报错
function f(a, ...b, c) {
  // ...
}
```

3.3.3　name 属性

调用函数的 name 属性会返回该函数的函数名：

```
function foo() {}
foo.name // "foo"
```

ES6 对 name 属性的行为做出了一些修改，如果将一个匿名函数赋值给一个变量，ES5 的 name 属性会返回空字符串，而 ES6 的 name 属性会返回实际的函数名：

```
var f = function () {};

// ES5
f.name // ""
```

```
// ES6
f.name // "f"
```

上述代码中，变量 f 等于一个匿名函数，ES5 和 ES6 的 name 属性返回的值不一样。如果将一个具名函数赋值给一个变量，则 ES5 和 ES6 的 name 属性都返回这个具名函数原本的名字。

```
const bar = function baz() {};

// ES5
bar.name // "baz"

// ES6
bar.name // "baz"
```

Function 构造函数返回的函数实例，name 属性的值为 anonymous：

```
(new Function).name // "anonymous"
```

bind 返回的函数，name 属性值会加上 bound 前缀：

```
function foo() {};
foo.bind({}).name              // "bound foo"

(function(){}).bind({}).name   // "bound "
```

3.3.4 箭头函数

箭头函数提供了一种更加简洁的函数书写方式。其基本语法是：

```
参数 => 函数体
```

例如：

```
var f = v => v;
//等价于
var f = function(a){
 return a;
}
f(1); //1
```

当箭头函数没有参数或者有多个参数时，要使用()：

```
var f = (a,b) => a+b;
f(6,2); //8
```

当箭头函数函数体有多行语句时，用{}包裹起来，表示代码块；当只有一行语句并且需要返回结果时，可以省略{}，结果会自动返回。例如：

```
var f = (a,b) => {
 let result = a+b;
 return result;
}
f(6,2);  // 8
```

当箭头函数要返回对象的时候,为了区分于代码块,要用()将对象包裹:

```
// 报错
var f = (id,name) => {id: id, name: name};
f(6,2);  // SyntaxError: Unexpected token :

// 不报错
var f = (id,name) => ({id: id, name: name});
f(6,2);  // {id: 6, name: 2}
```

【示例 3-9】需要注意的是,箭头函数内部没有 this、super、arguments 和 new.target 的引用。

```
var func = () => {
  // 箭头函数里面没有 this 对象
  // 此时的 this 是外层的 this 对象,即 Window
  console.log(this)
}
func(55)  // Window

var func = () => {
  console.log(arguments)
}
func(55);  // ReferenceError: arguments is not defined
```

【示例 3-10】箭头函数体中的 this 对象是定义函数时的对象,而不是使用函数时的对象。

```
function fn(){
  setTimeout(()=>{
    // 定义时,this 绑定的是 fn 中的 this 对象
    console.log(this.a);
  },0)
}
var a = 20;
// fn 的 this 对象为 {a: 19}
fn.call({a: 18});  // 18
```

不可以作为构造函数,也就是不能使用 new 命令,否则会报错。

【示例 3-11】ES6 之前,JavaScript 的 this 对象一直很复杂,回调函数中为了将外部 this 传递到回调函数中,经常看到 var self = this 的写法。有了箭头函数之后,就不需要这样做了,直接使用 this 即可:

```
// 回调函数
var Person = {
    'age': 18,
    'sayHello': function () {
      setTimeout(function () {
        console.log(this.age);
      });
    }
};
var age = 20;
Person.sayHello();  // 20

var Person1 = {
    'age': 18,
    'sayHello': function () {
      setTimeout(()=>{
        console.log(this.age);
      });
    }
};
var age = 20;
Person1.sayHello();  // 18
```

因此，当需要维护一个 this 上下文时，可以使用箭头函数进行处理。

【示例 3-12】若定义函数的方法，且该方法中包含 this，则不适合使用箭头函数。例如：

```
var Person = {
    'age': 18,
    'sayHello': ()=>{
        console.log(this.age);
    }
};
var age = 20;
Person.sayHello();  // 20
// 此时 this 指向的是全局对象

var Person1 = {
    'age': 18,
    'sayHello': function () {
        console.log(this.age);
    }
};
var age = 20;
Person1.sayHello();  // 18
// 此时 this 指向 Person1 对象
```

需要动态 this 的时候，也不适合使用箭头函数：

```
var button = document.getElementById('userClick');
button.addEventListener('click', () => {
    this.classList.toggle('on');
});
```

button 的监听函数是箭头函数，所以监听函数里面的 this 指向的是定义的时候外层的 this 对象，即 Window，导致无法操作到被单击的按钮对象。

3.4 闭包

众所周知，JavaScript 的一大特色就是闭包的存在，能够让内部函数访问到外部函数的变量。那么，我们该怎样认识闭包，以及在 ES6 标准下闭包会是怎样的呢？

【示例 3-13】闭包代码：

```
var test = function () {
  var arr = []
  for(var i = 0; i < 5; i++){
    arr.push(function () {
      return i*i
    })
  }
  return arr
}

var test1 = test()
console.log(test1[0]())
console.log(test1[1]())
console.log(test1[2]())
```

打印出来的 3 个值均为 25。在这个 for 循环内部为数组 push 了 5 个函数，代码为：

```
function () {
  return i*i
}
```

我们期望这里内部函数 i 能够获取到每次循环的 i 值，执行 test 函数，并赋给 test1。这样 test1 就成为了一个数组，并且使其内部元素均为函数。需要注意，这里内部的函数都还未执行，它们还没有在任何地方被调用。

在打印这一步时能够清楚地看到，开始调用其内部的函数。注意调用的时间是在打印这一步，那么此时的 i 值为多少呢？显然，此时已经在循环外部，i 值为 5。所以这也是打印值均为 25 的原因。

在 JavaScript 中，不存在块级作用域（先不讨论 ES6），只有函数作用域。作用域的好处是在内部函数可以访问外部函数的变量。

【示例 3-14】我们来改造一下这段代码：

```
var test = function () {
  var arr = []
  for(var i = 0; i < 5; i++){
    arr.push(function (n) {
      return n*n
    }(i))
  }
  return arr
}

var test1 = test()
console.log(test1)
```

这样循环每次执行时，将 i 作为参数实时传到内部函数中，这样便实现了保存每次 i 的值。但是和上面的函数不一样的地方是，我们这里获得的 test1 实际是一个由数字组成的数组，而不是函数，这和我们的目标是不符的。

【示例 3-15】我们需要的是数组内每个函数都能保存其创建时的 i 值，那么简单，在之前立即执行函数的基础上，让它的返回值也成为一个函数就行，而参数既然能往下传一层，当然也可以传两层，修改上述代码：

```
var test = function () {
  var arr = []
  for(var i = 0; i < 5; i++){
    arr.push(function (n) {
      return function () {
        return n * n
      }
    }(i))
  }
  return arr
}

var test1 = test()
console.log(test1[0]())
console.log(test1[1]())
console.log(test1[2]())
```

结果为：0、1、4。

可以看到，我们只是在之前的基础上进行了一点加工，让外层匿名函数的返回值也变为一个函数，让内层函数的返回值成为我们需要的结果，由于内层函数保持着对传进来的参数 i 的引用，导致 i 的值一直保存在内存中没有被释放，使得我们在调用这个函数的时候仍然能获得每次 i 的值，这也就是我们所说的闭包的实现。

在 ES6 中获得了两个新的声明方式：let 和 const。let 允许声明一个作用域被限制在块级中的变量、语句或者表达式。与 var 关键字不同的是，var 声明的变量只能是全局或者整个函数块的。

【示例 3-16】将上述代码使用 let 进行修改：

```
var test = function () {
  const arr = []
  for(let = 0; i < 5; i++){
    arr.push(function () {
      return i*i
    })
  }
  return arr
}

var test1 = test()
console.log(test1[0]())
console.log(test1[1]())
console.log(test1[2]())
```

相比于初始代码，整个函数有两处被替换：const arr = [] 以及 let i = 0，由于 let 声明的块级作用域，每次循环的 i 都会被直接固定下来而不会受其他地方的影响，轻松实现了闭包的效果。

所以在 ES6 环境下，我们不需要用立即执行函数的形式来实现闭包，使用 let 声明块级作用域即可。

3.5 对象

ES6 对对象进行了很多扩展，如属性和方法的简写方式、Object.assign()等。本节介绍 ES6 中 Object 对象的新增方法。

- 属性的简洁表示
- 属性名表达式
- 方法的 name 属性
- Object.is()

- Object.assign()
- 属性的可枚举性：Object.getOwnPropertyDescriptors()
- __proto__属性、Object.setPrototypeOf()、Object.getPrototypeOf()
- Object.keys()、Object.values()、Object.entries()
- Object.fromEntries()
- 对象的扩展运算符

3.5.1 属性的简洁表示

ES6 允许直接写入变量和函数，作为对象的属性和方法，这时属性名是变量名，属性值是变量值：

```
var foo = 'bar';
var baz = {foo};
baz // {foo: "bar"}

// 等同于
var baz = {foo: foo};
```

例如：

```
function f(x, y) {
  return {x, y};
}

// 等同于

function f(x, y) {
  return {x: x, y: y};
}

f(1, 2) // Object {x: 1, y: 2}
```

除了属性简写外，方法也可以简写：

```
var o = {
  method() {
    return "Hello!";
  }
};

// 等同于

var o = {
  method: function() {
    return "Hello!";
```

```
    }
};
```

【示例 3-17】变量属性使用示例：
```
var birth = '2000/01/01';

var Person = {

  name: '张三',

  //等同于 birth: birth
  birth,

  // 等同于 hello: function ()...
  hello() { console.log('我的名字是', this.name); }

};
```

这种写法用于函数的返回值，将会非常方便。

【示例 3-18】如果某个方法的值是一个 Generator 函数，前面需要加上星号。
```
const obj = {
  * myGenerator() {
    yield 'hello world';
  }
};
//等同于
const obj = {
  myGenerator: function* () {
    yield 'hello world';
  }
};
```

3.5.2 属性名表达式

JavaScript 语言定义对象的属性，有两种方法：
```
// 方法一
obj.foo = true;

// 方法二
obj['a' + 'bc'] = 123;
```

上面代码的方法一是直接用标识符作为属性名，方法二是用表达式作为属性名，这时要将表达式放在方括号内。但是，如果使用字面量方式定义对象（使用大括号），在 ES5 中只能使用方法一（标识符）定义属性。

```
var obj = {
  foo: true,
  abc: 123
};
```

ES6 允许字面量定义对象时使用方法二（表达式）作为对象的属性名，即把表达式放在方括号内，例如：

```
let propKey = 'foo';

let obj = {
  [propKey]: true,
  ['a' + 'bc']: 123
};
```

类似地，表达式还可以用于定义方法名：

```
let obj = {
  ['h' + 'ello']() {
    return 'hi';
  }
};

obj.hello() // hi
```

注意，属性名表达式与简洁表示法不能同时使用，否则会报错：

```
// 报错
var foo = 'bar';
var bar = 'abc';
var baz = { [foo] };

// 正确
var foo = 'bar';
var baz = { [foo]: 'abc'};
```

注意，属性名表达式如果是一个对象，默认情况下就会自动将对象转为字符串[object Object]，这一点要特别注意：

```
const keyA = {a: 1};
const keyB = {b: 2};

const myObject = {
  [keyA]: 'valueA',
  [keyB]: 'valueB'
};
```

```
myObject // Object {[object Object]: "valueB"}
```

上面的代码中，[keyA]和[keyB]得到的都是[object Object]，所以[keyB]会把[keyA]覆盖掉，而 myObject 最后只有一个[object Object]属性。

3.5.3 方法的 name 属性

函数的 name 属性返回函数名。对象方法也是函数，因此也有 name 属性，例如：

```
var person = {
  sayName() {
    console.log(this.name);
  },
  get firstName() {
    return "Nicholas";
  }
};

person.sayName.name       // "sayName"
person.firstName.name     // "get firstName"
```

上面的代码中，方法的 name 属性返回函数名（方法名）。如果使用了取值函数，就会在方法名的前面加上 get。如果是存值函数，方法名的前面就会加上 set。

有两种特殊情况，bind 方法创造的函数，name 属性返回"bound"加上原函数的名字；Function 构造函数创造的函数，name 属性返回"anonymous"。

```
(new Function()).name // "anonymous"

var doSomething = function() {
  // ...
};
doSomething.bind().name // "bound doSomething"
```

如果对象的方法是一个 Symbol 值，那么 name 属性返回的是这个 Symbol 值的描述。

```
const key1 = Symbol('description');
const key2 = Symbol();
let obj = {
  [key1]() {},
  [key2]() {},
};
obj[key1].name // "[description]"
obj[key2].name // ""
```

上面的代码中，key1 对应的 Symbol 值有描述，key2 没有描述。

3.5.4 对象的扩展运算符

对象的扩展运算符即将 Rest 运算符（解构赋值）/扩展运算符（...）引入对象，Babel 转码器已经支持这项功能。

对象的解构赋值用于从一个对象取值，相当于将所有可遍历的但尚未被读取的属性分配到指定的对象上面。所有的键和它们的值都会拷贝到新对象上面，例如：

```
let { x, y, ...z } = { x: 1, y: 2, a: 3, b: 4 };
x // 1
y // 2
z // { a: 3, b: 4 }
```

上面的代码中，变量 z 是解构赋值所在的对象。它获取等号右边的所有尚未读取的键（a 和 b），将它们连同值一起拷贝过来。

由于解构赋值要求等号右边是一个对象，因此如果等号右边是 undefined 或 null，就会报错，因为它们无法转为对象。

```
let { x, y, ...z } = null;            // 运行时错误
let { x, y, ...z } = undefined;       // 运行时错误
```

解构赋值必须是最后一个参数，否则会报错：

```
let { ...x, y, z } = obj;             // 句法错误
let { x, ...y, ...z } = obj;          // 句法错误
```

上面的代码中，解构赋值不是最后一个参数，所以会报错。

注意，解构赋值的拷贝是浅拷贝，即如果一个键的值是复合类型的值（数组、对象、函数），那么解构赋值拷贝的是这个值的引用，而不是这个值的副本。

```
let obj = { a: { b: 1 } };
let { ...x } = obj;
obj.a.b = 2;
x.a.b // 2
```

上面的代码中，x 是解构赋值所在的对象，拷贝了对象 obj 的 a 属性。a 属性引用了一个对象，修改这个对象的值会影响解构赋值对它的引用。同时，解构赋值不会拷贝继承自原型对象的属性：

```
let o1 = { a: 1 };
let o2 = { b: 2 };
o2.__proto__ = o1;
let o3 = { ...o2 };
o3 // { b: 2 }
```

上面的代码中，对象 o3 是 o2 的拷贝，但是只复制了 o2 自身的属性，没有复制它的原型对象 o1 的属性。

扩展运算符（...）用于取出参数对象的所有可遍历属性，拷贝到当前对象中，例如：

```
let person = {name: "Amy", age: 15};
let someone = { ...person };
someone;  //{name: "Amy", age: 15}
```

这等同于使用 Object.assign 方法：

```
let aClone = { ...a };
// 等同于
let aClone = Object.assign({}, a);
```

扩展运算符可以用于合并两个对象，例如：

```
let age = {age: 15};
let name = {name: "Amy"};
let person = {...age, ...name};
person;  //{age: 15, name: "Amy"}
```

如果用户自定义的属性放在扩展运算符后面，扩展运算符内部的同名属性就会被覆盖掉。

```
let aWithOverrides = { ...a, x: 1, y: 2 };
// 等同于
let aWithOverrides = { ...a, ...{ x: 1, y: 2 } };
// 等同于
let x = 1, y = 2, aWithOverrides = { ...a, x, y };
// 等同于
let aWithOverrides = Object.assign({}, a, { x: 1, y: 2 });
```

上面的代码中，a 对象的 x 属性和 y 属性拷贝到新对象后会被覆盖掉。这用来修改现有对象的部分属性就很方便。

```
let newVersion = {
  ...previousVersion,
  name: 'New Name' // Override the name property
};
```

上面的代码中，newVersion 对象自定义了 name 属性，其他属性全部复制自 previousVersion 对象。

如果把自定义属性放在扩展运算符前面，就变成了设置新对象的默认属性值：

```
let aWithDefaults = { x: 1, y: 2, ...a };
// 等同于
let aWithDefaults = Object.assign({}, { x: 1, y: 2 }, a);
// 等同于
let aWithDefaults = Object.assign({ x: 1, y: 2 }, a);
```

3.5.5 对象的新方法

1. Object.is()

ES5 比较两个值是否相等只有两个运算符：相等运算符（==）和严格相等运算符（===）。它们都有缺点，前者会自动转换数据类型，后者的 NaN 不等于自身，以及+0 等于-0。JavaScript 不支持这种运算，在所有环境中，只要两个值是一样的，它们就应该相等。

ES6 提出 Same-Value Equality（同值相等）算法，用来解决这个问题。Object.is 就是部署这个算法的新方法。它用来比较两个值是否严格相等，与严格比较运算符（===）的行为基本一致。

```
Object.is('foo', 'foo')
// true
Object.is({}, {})
// false
```

不同之处只有两个：一是+0 不等于-0，二是 NaN 等于自身。

```
+0 === -0           //true
NaN === NaN         // false

Object.is(+0, -0)    // false
Object.is(NaN, NaN)  // true
```

ES5 可以通过下面的代码部署 Object.is。

```
Object.defineProperty(Object, 'is', {
  value: function(x, y) {
    if (x === y) {
      // 针对+0 不等于 -0的情况
      return x !== 0 || 1 / x === 1 / y;
    }
    // 针对 NaN 的情况
    return x !== x && y !== y;
  },
  configurable: true,
  enumerable: false,
  writable: true
});
```

2. Object.assign()

Object.assign 方法用于对象的合并，将源对象（source）的所有可枚举属性复制到目标对象（target）。

```
var target = { a: 1 };

var source1 = { b: 2 };
```

```
var source2 = { c: 3 };

Object.assign(target, source1, source2);
target // {a:1, b:2, c:3}
```

Object.assign 方法的第一个参数是目标对象,后面的参数都是源对象。

注意,如果目标对象与源对象有同名属性,或多个源对象有同名属性,后面的属性就会覆盖前面的属性。

```
var target = { a: 1, b: 1 };

var source1 = { b: 2, c: 2 };
var source2 = { c: 3 };

Object.assign(target, source1, source2);
target // {a:1, b:2, c:3}
```

如果只有一个参数,Object.assign 会直接返回该参数。

```
var obj = {a: 1};
Object.assign(obj) === obj  // true
```

如果该参数不是对象,就会先转成对象,然后返回。

```
typeof Object.assign(2)     // "object"
```

由于 undefined 和 null 无法转成对象,因此如果它们作为参数,就会报错。

```
Object.assign(undefined)    // 报错
Object.assign(null)         // 报错
```

如果非对象参数出现在源对象的位置(非首参数),那么处理规则有所不同。首先,这些参数都会转成对象,如果无法转成对象,就会跳过。这意味着,如果 undefined 和 null 不在首参数,就不会报错。

```
let obj = {a: 1};
Object.assign(obj, undefined) === obj    // true
Object.assign(obj, null) === obj         // true
```

其他类型的值(数值、字符串和布尔值)不在首参数,也不会报错。但是,除了字符串会以数组形式拷贝入目标对象外,其他值都不会产生效果。

```
var v1 = 'abc';
var v2 = true;
var v3 = 10;

var obj = Object.assign({}, v1, v2, v3);
console.log(obj); // { "0": "a", "1": "b", "2": "c" }
```

上面的代码中，v1、v2、v3 分别是字符串、布尔值和数值，结果只有字符串可以合并到目标对象中（以字符数组的形式），数值和布尔值都会被忽略。这是因为只有字符串的包装对象会产生可枚举属性。

```
Object(true) // {[[PrimitiveValue]]: true}
Object(10)   // {[[PrimitiveValue]]: 10}
Object('abc') // {0: "a", 1: "b", 2: "c", length: 3, [[PrimitiveValue]]: "abc"}
```

上面的代码中，布尔值、数值、字符串分别转成对应的包装对象，可以看到它们的原始值都在包装对象的内部属性[[PrimitiveValue]]上面，这个属性是不会被 Object.assign 拷贝的。只有字符串的包装对象会产生可枚举的实义属性，这些属性才会被拷贝。

Object.assign 拷贝的属性是有限制的，只拷贝源对象的自身属性，不拷贝继承属性，也不拷贝不可枚举的属性（enumerable: false）。

```
Object.assign({b: 'c'},
  Object.defineProperty({}, 'invisible', {
    enumerable: false,
    value: 'hello'
  })
)
// { b: 'c' }
```

上面的代码中，Object.assign 要拷贝的对象只有一个不可枚举属性 invisible，这个属性并没有被拷贝进去。

属性名为 Symbol 值的属性也会被 Object.assign 拷贝。

```
Object.assign({ a: 'b' }, { [Symbol('c')]: 'd' })
// { a: 'b', Symbol(c): 'd' }
```

注意：Object.assign 方法实行的是浅拷贝，而不是深拷贝。也就是说，如果源对象某个属性的值是对象，那么目标对象拷贝得到的是这个对象的引用。

```
var obj1 = {a: {b: 1}};
var obj2 = Object.assign({}, obj1);

obj1.a.b = 2;
obj2.a.b // 2
```

上面的代码中，源对象 obj1 的 a 属性的值是一个对象，Object.assign 拷贝得到的是这个对象的引用。这个对象的任何变化都会反映到目标对象上面。

对于这种嵌套的对象，一旦遇到同名属性，Object.assign 的处理方法是替换，而不是添加。

```
var target = { a: { b: 'c', d: 'e' } }
var source = { a: { b: 'hello' } }
Object.assign(target, source)
// { a: { b: 'hello' } }
```

上面的代码中，target 对象的 a 属性被 source 对象的 a 属性整个替换掉了，而不会得到{ a: { b: 'hello', d: 'e' } }的结果。这通常不是开发者想要的，需要特别小心。

有一些函数库提供 Object.assign 的定制版本（比如 Lodash 的_.defaultsDeep 方法），可以解决浅拷贝的问题，得到深拷贝的合并。

注意，Object.assign 可以用来处理数组，但是会把数组视为对象。

```
Object.assign([1, 2, 3], [4, 5])
// [4, 5, 3]
```

上面的代码中，Object.assign 把数组视为属性名为 0、1、2 的对象，因此目标数组的 0 号属性 4 覆盖了原数组的 0 号属性 1。

Object.assign 方法有很多用处，下面分别说明。

（1）为对象添加属性

```
class Point {
  constructor(x, y) {
    Object.assign(this, {x, y});
  }
}
```

上面的通过 Object.assign 方法将 x 属性和 y 属性添加到 Point 类的对象实例。

（2）为对象添加方法

```
Object.assign(SomeClass.prototype, {
  someMethod(arg1, arg2) {
    ...
  },
  anotherMethod() {
    ...
  }
});

// 等同于下面的写法
SomeClass.prototype.someMethod = function (arg1, arg2) {
  ...
};
SomeClass.prototype.anotherMethod = function () {
  ...
};
```

上面的代码使用了对象属性的简洁表示法，直接将两个函数放在大括号中，再使用 assign 方法添加到 SomeClass.prototype 中。

（3）克隆对象

```
function clone(origin) {
  return Object.assign({}, origin);
}
```

上面的代码将原始对象拷贝到一个空对象，就得到了原始对象的克隆。

不过，采用这种方法克隆只能克隆原始对象自身的值，不能克隆它继承的值。如果想要保持继承链，可以采用下面的代码。

```
function clone(origin) {
  let originProto = Object.getPrototypeOf(origin);
  return Object.assign(Object.create(originProto), origin);
}
```

（4）合并多个对象

将多个对象合并到某个对象。

```
const merge =
  (target, ...sources) => Object.assign(target, ...sources);
```

如果希望合并后返回一个新对象，可以改写上面的函数，对一个空对象合并：

```
const merge =
  (...sources) => Object.assign({}, ...sources);
```

（5）为属性指定默认值

```
const DEFAULTS = {
  logLevel: 0,
  outputFormat: 'html'
};

function processContent(options) {
  options = Object.assign({}, DEFAULTS, options);
}
```

上面的代码中，DEFAULTS 对象是默认值，options 对象是用户提供的参数。Object.assign 方法将 DEFAULTS 和 options 合并成一个新对象，如果两者有同名属性，option 的属性值就会覆盖 DEFAULTS 的属性值。

> **注　意**
>
> 由于存在深拷贝的问题，DEFAULTS 对象和 options 对象的所有属性的值都只能是简单类型，而不能指向另一个对象。否则，将导致 DEFAULTS 对象的该属性不起作用。

3. Object.keys()

ES5 引入了 Object.keys 方法，返回一个数组，成员是参数对象自身的（不含继承的）所有可遍历（Enumerable）属性的键名。

```
var obj = { foo: "bar", baz: 42 };
Object.keys(obj)
// ["foo", "baz"]
```

目前，ES7 有一个提案，引入了跟 Object.keys 配套的 Object.values 和 Object.entries。

```
let {keys, values, entries} = Object;
let obj = { a: 1, b: 2, c: 3 };

for (let key of keys(obj)) {
  console.log(key); // 'a', 'b', 'c'
}

for (let value of values(obj)) {
  console.log(value); // 1, 2, 3
}

for (let [key, value] of entries(obj)) {
  console.log([key, value]); // ['a', 1], ['b', 2], ['c', 3]
}
Object.values()
```

Object.values 方法返回一个数组，成员是参数对象自身的（不含继承的）所有可遍历属性的键值。

```
var obj = { foo: "bar", baz: 42 };
Object.values(obj)
// ["bar", 42]
```

返回数组的成员顺序与本章的 3.5.7 小节介绍的排列规则一致。

```
var obj = { 100: 'a', 2: 'b', 7: 'c' };
Object.values(obj)
// ["b", "c", "a"]
```

上面的代码中，属性名为数值的属性，是按照数值大小从小到大遍历的，因此返回的顺序是 b、c、a。

Object.values 只返回对象自身的可遍历属性。

```
var obj = Object.create({}, {p: {value: 42}});
Object.values(obj) // []
```

上面的代码中，Object.create 方法的第二个参数添加的对象属性（属性 p），如果不显式声明，默认是不可遍历的。Object.values 不会返回这个属性。

Object.values 会过滤属性名为 Symbol 值的属性。

```
Object.values({ [Symbol()]: 123, foo: 'abc' });
// ['abc']
```

如果 Object.values 方法的参数是一个字符串，就会返回各个字符组成的一个数组。

```
Object.values('foo')
// ['f', 'o', 'o']
```

上面的代码中，字符串会先转成一个类似数组的对象。字符串的每个字符就是该对象的一个属性。因此，Object.values 返回每个属性的键值，就是各个字符组成的一个数组。

如果参数不是对象，Object.values 就会先将其转为对象。由于数值和布尔值的包装对象都不会为实例添加非继承的属性，因此 Object.values 会返回空数组。

```
Object.values(42) // []
Object.values(true) // []
```

4. Object.entries()

Object.entries 方法返回一个数组，成员是参数对象自身的（不含继承的）所有可遍历属性的键值对数组。

```
var obj = { foo: 'bar', baz: 42 };
Object.entries(obj)
// [ ["foo", "bar"], ["baz", 42] ]
```

除了返回值不一样外，该方法的行为与 Object.values 基本一致。

如果原对象的属性名是一个 Symbol 值，该属性就会被省略。

```
Object.entries({ [Symbol()]: 123, foo: 'abc' });
// [ [ 'foo', 'abc' ] ]
```

上面的代码中，原对象有两个属性，Object.entries 只输出属性名非 Symbol 值的属性。将来可能会有 Reflect.ownEntries()方法，返回对象自身的所有属性。

Object.entries 的基本用途是遍历对象的属性。

```
let obj = { one: 1, two: 2 };
for (let [k, v] of Object.entries(obj)) {
  console.log(`${JSON.stringify(k)}: ${JSON.stringify(v)}`);
}
// "one": 1
// "two": 2
```

Object.entries 方法的一个用处是将对象转为真正的 Map 结构。

```
var obj = { foo: 'bar', baz: 42 };
var map = new Map(Object.entries(obj));
map // Map { foo: "bar", baz: 42 }
```

自己实现 Object.entries 方法，非常简单。

```
// Generator 函数的版本
function* entries(obj) {
  for (let key of Object.keys(obj)) {
    yield [key, obj[key]];
  }
}

// 非 Generator 函数的版本
function entries(obj) {
  let arr = [];
  for (let key of Object.keys(obj)) {
    arr.push([key, obj[key]]);
  }
  return arr;
}
```

5. __proto__ 属性、Object.setPrototypeOf()和 Object.getPrototypeOf()

（1）__proto__ 属性

__proto__ 属性（前后各两个下划线）用来读取或设置当前对象的 prototype 对象。目前，所有浏览器（包括 IE 11）都部署了这个属性。

```
// ES6的写法
var obj = {
  method: function() { ... }
};
obj.__proto__ = someOtherObj;

// es5的写法
var obj = Object.create(someOtherObj);
obj.method = function() { ... };
```

该属性没有写入 ES6 的正文，而是写入了附录，原因是 __proto__ 前后的双下划线说明它本质上是一个内部属性，而不是一个正式的、对外的 API，只是由于浏览器的广泛支持，才被加入了 ES6。标准明确规定，浏览器必须部署这个属性，其他运行环境不一定需要部署，而且新的代码最好认为这个属性是不存在的。因此，无论从语义的角度，还是从兼容性的角度，都不要使用这个属性，而是使用 Object.setPrototypeOf()（写操作）、Object.getPrototypeOf()（读操作）、Object.create()（生成操作）代替。

在实现上，__proto__ 调用的是 Object.prototype.__proto__，具体实现如下：

```
Object.defineProperty(Object.prototype, '__proto__', {
  get() {
    let _thisObj = Object(this);
    return Object.getPrototypeOf(_thisObj);
  },
```

```
    set(proto) {
      if (this === undefined || this === null) {
        throw new TypeError();
      }
      if (!isObject(this)) {
        return undefined;
      }
      if (!isObject(proto)) {
        return undefined;
      }
      let status = Reflect.setPrototypeOf(this, proto);
      if (!status) {
        throw new TypeError();
      }
    },
});
function isObject(value) {
  return Object(value) === value;
}
```

如果一个对象本身部署了__proto__属性，该属性的值就是对象的原型。

```
Object.getPrototypeOf({ __proto__ : null })
// null
```

（2）Object.setPrototypeOf()

Object.setPrototypeOf 方法的作用与__proto__相同，用来设置一个对象的 prototype 对象。它是 ES6 正式推荐的设置原型对象的方法。

```
// 格式
Object.setPrototypeOf(object, prototype)

// 用法
var o = Object.setPrototypeOf({}, null);
```

该方法等同于下面的函数。

```
function (obj, proto) {
  obj.__proto__ = proto;
  return obj;
}
```

下面是一个例子。

```
let proto = {};
let obj = { x: 10 };
Object.setPrototypeOf(obj, proto);
```

```
proto.y = 20;
proto.z = 40;

obj.x // 10
obj.y // 20
obj.z // 40
```

上面的代码将 proto 对象设为 obj 对象的原型，所以从 obj 对象可以读取 proto 对象的属性。

（3）Object.getPrototypeOf()

该方法与 setPrototypeOf 方法配套，用于读取一个对象的 prototype 对象。

```
Object.getPrototypeOf(obj);
```

下面是一个例子。

```
function Rectangle() {
}

var rec = new Rectangle();

Object.getPrototypeOf(rec) === Rectangle.prototype
// true

Object.setPrototypeOf(rec, Object.prototype);
Object.getPrototypeOf(rec) === Rectangle.prototype
// false
```

3.5.6 属性的可枚举性

对象的每个属性都有一个描述对象（Descriptor），用来控制该属性的行为。Object.getOwnPropertyDescriptor 方法可以获取该属性的描述对象。

```
let obj = { foo: 123 };
Object.getOwnPropertyDescriptor(obj, 'foo')
// {
//   value: 123,
//   writable: true,
//   enumerable: true,
//   configurable: true
// }
```

描述对象的 enumerable 属性称为"可枚举性"，如果该属性为 false，就表示某些操作会忽略当前属性。

ES5 有 3 个操作会忽略 enumerable 为 false 的属性。

- for...in 循环：只遍历对象自身的和继承的可枚举的属性。

- Object.keys()：返回对象自身的所有可枚举的属性的键名。
- JSON.stringify()：只串行化对象自身的可枚举的属性。

ES6 新增了一个操作 Object.assign()，会忽略 enumerable 为 false 的属性，只拷贝对象自身可枚举的属性。

这 4 个操作中，只有 for...in 会返回继承的属性。实际上，引入 enumerable 的最初目的就是让某些属性可以规避掉 for...in 操作。比如，对象原型的 toString 方法以及数组的 length 属性就通过这种手段避免被 for...in 遍历到。

```
Object.getOwnPropertyDescriptor(Object.prototype, 'toString').enumerable
// false

Object.getOwnPropertyDescriptor([], 'length').enumerable
// false
```

上面的代码中，toString 和 length 属性的 enumerable 都是 false，因此 for...in 不会遍历到这两个继承自原型的属性。

另外，ES6 规定，所有 Class 的原型的方法都是不可枚举的。

```
Object.getOwnPropertyDescriptor(class {foo() {}}.prototype, 'foo').enumerable
// false
```

总的来说，操作中引入继承的属性会让问题复杂化，大多数情况下，我们只关心对象自身的属性。所以，尽量不要用 for...in 循环，而用 Object.keys()代替。

3.5.7　属性的遍历

ES6 一共有 5 种方法可以遍历对象的属性。

（1）for...in

for...in 循环遍历对象自身的和继承的可枚举属性（不含 Symbol 属性）。

（2）Object.keys(obj)

Object.keys 返回一个数组，包括对象自身的（不含继承的）所有可枚举属性（不含 Symbol 属性）。

（3）Object.getOwnPropertyNames(obj)

Object.getOwnPropertyNames 返回一个数组，包含对象自身的所有属性（不含 Symbol 属性，但是包括不可枚举属性）。

（4）Object.getOwnPropertySymbols(obj)

Object.getOwnPropertySymbols 返回一个数组，包含对象自身的所有 Symbol 属性。

（5）Reflect.ownKeys(obj)

Reflect.ownKeys 返回一个数组，包含对象自身的所有属性，不管属性名是 Symbol 或字符串，也不管是否可枚举。

以上的 5 种方法遍历对象的属性遵守同样的属性遍历的次序规则：
- 首先遍历所有属性名为数值的属性，按照数字排序。
- 其次遍历所有属性名为字符串的属性，按照生成时间排序。
- 最后遍历所有属性名为 Symbol 值的属性，按照生成时间排序。

```
Reflect.ownKeys({ [Symbol()]:0, b:0, 10:0, 2:0, a:0 })
// ['2', '10', 'b', 'a', Symbol()]
```

上面的代码中，Reflect.ownKeys 方法返回一个数组，包含参数对象的所有属性。这个数组的属性次序是这样的，首先是数值属性 2 和 10，其次是字符串属性 b 和 a，最后是 Symbol 属性。

3.6 类

严格意义上来讲，Node.js 的类不能算是类，其实它只是一个函数的集合体，加一些成员变量。不过接下来我们都将其称为"类"，实例化的叫"对象"。因为类有着很多函数的特性，或者说类的本质是一个函数。本节介绍类的静态方法、原型方法、实例方法、实例化、类修饰器以及类方法修饰器、类的继承等。

3.6.1 基础用法

在 ES6 中，class（类）作为对象的模板被引入，可以通过 class 关键字定义类。class 的本质是 function。在 ES6 中，class 可以看作一个语法糖，让对象原型的写法更加清晰，更像面向对象编程的语法。

【示例 3-19】类的定义：

```
// 匿名类
let Example = class {
    constructor(a) {
        this.a = a;
    }
}
// 命名类
let Example = class Example {
    constructor(a) {
        this.a = a;
```

```
    }
}
```

从示例中可以看出类表达式可以为匿名或命名，匿名即为 class {}的部分，命名即为 class Example {}的部分。

【示例3-20】类声明：

```
class Example {
    constructor(a) {
        this.a = a;
    }
}
```

需要注意的是，不可重复声明同一个类，声明同一个类将报出异常信息：

```
class Example{}
class Example{}
// Uncaught SyntaxError: Identifier 'Example' has already been
// declared

let Example1 = class{}
class Example{}
// Uncaught SyntaxError: Identifier 'Example' has already been
// declared
```

> **注 意**
>
> 类定义不会被提升，这意味着必须在访问类之前对类进行定义，否则就会报错。

在 ES6 中，prototype 属性仍旧存在，虽然可以直接在类中定义方法，但是本质上类方法还是定义在 prototype 上的。例如覆盖 prototype 属性，在初始化时添加方法：

```
Example.prototype={
    //methods
}
```

添加方法：

```
Object.assign(Example.prototype,{
    //methods
})
```

将直接定义在类内部的属性（Class.propname）称为静态属性，静态属性不需要实例化。ES6 中规定，class 内部只有静态方法，没有静态属性。

```
class Example {
// 新提案
    static a = 2;
}
```

```
// 目前可行写法
Example.b = 2;
```

公共属性：

```
class Example{}
Example.prototype.a = 2;
```

实例属性是定义在实例对象（this）上的属性，例如：

```
class Example {
    a = 2;
    constructor () {
        console.log(this.a);
    }
}
```

Class 的 name 属性会返回跟在 class 关键字后的类名（命名方式定义类）：

```
let Example=class Exam {
    constructor(a) {
        this.a = a;
    }
}
console.log(Example.name); // Exam

let Example=class {
    constructor(a) {
        this.a = a;
    }
}
console.log(Example.name); // Example
```

类的 constructor 方法是类的默认方法，创建类的实例化对象时被调用：

```
class Example{
    constructor(){
      console.log('我是constructor');
    }
}
new Example(); // 我是constructor
```

constructor 默认返回实例对象 this，也可指定所需返回的对象：

```
class Test {
    constructor(){
        // 默认返回实例对象 this
    }
}
```

```
console.log(new Test() instanceof Test); // true

class Example {
    constructor(){
        // 指定返回对象
        return new Test();
    }
}
console.log(new Example() instanceof Example); // false
```

【示例 3-21】类的静态方法示例:

```
class Example{
    static sum(a, b) {
        console.log(a+b);
    }
}
Example.sum(1, 2); // 3
```

【示例 3-22】类的原型方法示例:

```
class Example {
    sum(a, b) {
        console.log(a + b);
    }
}
let exam = new Example();
exam.sum(1, 2); // 3
```

【示例 3-23】类的实例方法示例:

```
class Example {
    constructor() {
        this.sum = (a, b) => {
            console.log(a + b);
        }
    }
}
```

类的实例化必须通过 new 关键字,例如:

```
class Example {}

let exam1 = Example();
// Class constructor Example cannot be invoked without 'new'
```

【示例 3-24】实例化多个对象时共享原型对象:

```
class Example {
```

```
    constructor(a, b) {
        this.a = a;
        this.b = b;
        console.log('Example');
    }
    sum() {
        return this.a + this.b;
    }
}
let exam1 = new Example(2, 1);
let exam2 = new Example(3, 1);
console.log(exam1.__proto__ == exam2.__proto__); // true

exam1.__proto__.sub = function() {
    return this.a - this.b;
}
console.log(exam1.sub()); // 1
console.log(exam2.sub()); // 2
```

通过 decorator 类修饰器修改类的行为。decorator 是一个函数，在代码编译时产生作用。

【示例 3-25】decorator 的第一个参数 target 指向类本身：

```
function testable(target) {
    target.isTestable = true;
}
@testable
class Example {}
Example.isTestable; // true
```

【示例 3-26】或者通过嵌套实现修饰：

```
function testable(isTestable) {
    return function(target) {
        target.isTestable=isTestable;
    }
}
@testable(true)
class Example {}
Example.isTestable; // true
```

以上两个示例添加的是类的静态属性，若要添加实例属性，则需要在类的 prototype 上操作。

还可以对类的方法进行修饰。类方法修饰器接收 3 个参数：target（类的原型对象）、name（修饰的属性名）、descriptor（该属性的描述对象），例如：

```
class Example {
```

```
    @writable
    sum(a, b) {
        return a + b;
    }
}
function writable(target, name, descriptor) {
    descriptor.writable = false;
    return descriptor; // 必须返回
}
```

【示例 3-27】修饰器执行顺序示例：

```
class Example {
    @logMethod(1)
    @logMethod(2)
    sum(a, b){
        return a + b;
    }
}
function logMethod(id) {
    console.log('evaluated logMethod'+id);
    return (target, name, desctiptor) => console.log('excuted        logMethod'+id);
}
// evaluated logMethod 1
// evaluated logMethod 2
// excuted logMethod 2
// excuted logMethod 1
```

由上述示例可以看出，修饰器执行顺序是由外向内传入、由内向外执行的。

3.6.2 封装与继承

在"类"的内部可以使用 get 和 set 关键字，对某个属性设置存值函数和取值函数，拦截该属性的存取行为。

【示例 3-28】对某个属性设置存值函数和取值函数：

```
class Example{
    constructor(a, b) {
        this.a = a; // 实例化时调用 set 方法
        this.b = b;
    }
    get a(){
        console.log('getter');
        return this.a;
    }
```

```
    set a(a){
        console.log('setter');
        this.a = a;  // 自身递归调用
    }
}
let exam = new Example(1,2);  // 不断输出 setter，最终导致 RangeError
class Example1{
    constructor(a, b) {
        this.a = a;
        this.b = b;
    }
    get a(){
        console.log('getter');
        return this._a;
    }
    set a(a){
        console.log('setter');
        this._a = a;
    }
}
let exam1 = new Example1(1,2);      // 只输出 setter，不会调用 getter 方法
console.log(exam._a);               // 1，可以直接访问
```

上述代码中，a 属性有对应的存值函数和取值函数，因此赋值和读取行为都被自定义了。存值函数和取值函数是设置在属性的 Descriptor 对象上的。

注意，getter 不可单独出现，否则会报错：

```
class Example {
    constructor(a) {
        this.a = a;
    }
    get a() {
        return this.a;
    }
}
let exam = new Example(1); // Uncaught TypeError: Cannot set property // a of
#<Example> which has only a getter
```

getter 与 setter 必须同级出现：

```
class Father {
    constructor(){}
    get a() {
        return this._a;
    }
}
```

```
class Child extends Father {
    constructor(){
        super();
    }
    set a(a) {
        this._a = a;
    }
}
let test = new Child();
test.a = 2;
console.log(test.a); // undefined

class Father1 {
    constructor(){}
    // 或者都放在子类中
    get a() {
        return this._a;
    }
    set a(a) {
        this._a = a;
    }
}
class Child1 extends Father1 {
    constructor(){
        super();
    }
}
let test1 = new Child1();
test1.a = 2;
console.log(test1.a); // 2
```

Class 可以通过 extends 关键字实现类的继承，这比 ES5 通过修改原型链实现继承要清晰和方便很多，例如：

```
class Father {
}
class Child extends Father {
    ...
}
```

上面的代码定义了一个 Father 类，该类通过 extends 关键字继承了 Father 类的所有属性和方法。但是由于没有部署任何代码，因此这两个类完全一样，等于复制了一个 Father 类。后续我们在 Child 内部加上代码。

3.6.3　super 关键字

super 作为函数调用时，代表父类的构造函数。ES6 要求子类的构造函数必须执行一次 super 函数，即子类 constructor 方法中必须有 super，且必须出现在 this 之前。

【示例 3-29】

```
class Father {
    constructor() {}
}
class Child extends Father {
    constructor() {}
    // or
    // constructor(a) {
    //    this.a = a;
    //    super();
    // }
}
let test = new Child(); // Uncaught ReferenceError: Must call super
// constructor in derived class before accessing 'this' or returning
// from derived constructor
```

【示例 3-30】调用父类构造函数，只能出现在子类的构造函数中：

```
class Father {
    test(){
        return 0;
    }
    static test1(){
        return 1;
    }
}
class Child extends Father {
    constructor(){
        super();
    }
}
class Child1 extends Father {
    test2() {
        super(); // Uncaught SyntaxError: 'super' keyword unexpected
        // here
    }
}
```

【示例 3-31】调用父类方法，super 作为对象，在普通方法中，指向父类的原型对象，在静态方法中，指向父类：

```
class Child2 extends Father {
    constructor(){
        super();
        // 调用父类普通方法
        console.log(super.test()); // 0
    }
    static test3(){
        // 调用父类静态方法
        return super.test1+2;
    }
}
Child2.test3(); // 3
```

注意，不可继承常规对象：

```
var Father = {
    // ...
}
class Child extends Father {
    // ...
}
// Uncaught TypeError: Class extends value #<Object> is not a constructor or null

// 解决方案
Object.setPrototypeOf(Child.prototype, Father);
```

3.7　ES6 的模块化

在 ES6 之前，实现模块化使用的是 RequireJS 或者 seaJS（分别是基于 AMD 规范的模块化库和基于 CMD 规范的模块化库）。ES6 引入了模块化，其设计思想是在编译时就能确定模块的依赖关系，以及输入和输出的变量。ES6 的模块化分为导出（export）与导入（import）两个模块。

不论是否在模块头部加上 use strict;，ES6 的模块自动开启严格模式。模块中可以导入和导出各种类型的变量，如函数、对象、字符串、数字、布尔值、类等。每个模块都有自己的上下文，每一个模块内声明的变量都是局部变量，不会污染全局作用域。每一个模块只加载一次（是单例的），若再去加载同目录下的同文件，则直接从内存中读取。

3.7.1　基本用法

模块导入导出各种类型的变量，如字符串、数值、函数、类：

- 导出的函数声明与类声明必须有名称（export default 命令另外考虑）。
- 不仅能导出声明，还能导出引用（例如函数）。
- export 命令可以出现在模块的任何位置，但必须处于模块顶层。
- import 命令会提升到整个模块的头部，首先执行。

【示例 3-32】
```
/*-----export [test.js]-----*/
let myName = "Tom";
let myAge = 20;
let myfn = function(){
    return "My name is" + myName + "! I'm '" + myAge + "years old."
}
let myClass = class myClass {
    static a = "yeah!";
}
export { myName, myAge, myfn, myClass }

/*-----import [xxx.js]-----*/
import { myName, myAge, myfn, myClass } from "./test.js";
console.log(myfn());// My name is Tom! I'm 20 years old.
console.log(myAge);// 20
console.log(myName);// Tom
console.log(myClass.a );// yeah!
```

建议使用大括号指定所要输出的一组变量写在文档尾部，明确导出的接口。函数与类都需要有对应的名称，导出文档尾部也避免了无对应名称。

3.7.2　as 的用法

export 命令导出的接口名称必须和模块内部的变量有一一对应关系。导入的变量名必须和导出的接口名称相同，但顺序可以不一致。例如：

```
/*-----export [test.js]-----*/
let myName = "Tom";
export { myName as exportName }

/*-----import [xxx.js]-----*/
import { exportName } from "./test.js";
console.log(exportName);// Tom
```

【示例 3-33】使用 as 重新定义导出的接口名称，隐藏模块内部的变量：

```
/*-----export [test1.js]-----*/
let myName = "Tom";
export { myName }
```

```
/*-----export [test2.js]-----*/
let myName = "Jerry";
export { myName }
/*-----import [xxx.js]-----*/
import { myName as name1 } from "./test1.js";
import { myName as name2 } from "./test2.js";
console.log(name1);// Tom
console.log(name2);// Jerry
```

不同模块导出接口名称命名重复，即可使用 as 重新定义变量名。

3.7.3　import 命令的特点

（1）只读属性：不允许在加载模块的脚本里面改写接口的引用指向，即可以改写 import 变量类型为对象的属性值，不能改写 import 变量类型为基本类型的值，例如：

```
import {a} from "./xxx.js"
a = {}; // error

import {a} from "./xxx.js"
a.foo = "hello"; // a = { foo : 'hello' }
```

（2）单例模式：多次重复执行同一句 import 语句，那么只会执行一次，而不会执行多次。import 同一模块声明不同接口引用会声明对应变量，但只执行一次 import。

```
import { a } "./xxx.js";
import { a } "./xxx.js";
// 相当于 import { a } "./xxx.js";

import { a } from "./xxx.js";
import { b } from "./xxx.js";
// 相当于 import { a, b } from "./xxx.js";
```

（3）静态执行特性：import 是静态执行的，所以不能使用表达式和变量。

```
import { "f" + "oo" } from "methods";
// error
let module = "methods";
import { foo } from module;
// error
if (true) {
  import { foo } from "method1";
} else {
  import { foo } from "method2";
}
// error
```

(4) export default 命令的特点：

- 在一个文件或模块中，export、import 可以有多个，export default 仅有一个。
- export default 中的 default 是对应的导出接口变量。
- 通过 export 方式导出，在导入时要加{}，export default 则不需要。
- export default 向外暴露的成员可以使用任意变量来接收。

【示例 3-34】

```
var a = "My name is Tom!";
export default a; // 仅有一个
export default var c = "error";
// error, default 已经是对应的导出变量，不能跟着变量声明语句

import b from "./xxx.js"; // 不需要加{}, 使用任意变量接收
```

3.7.4　export 与 import

export 与 import 可以在同一模块使用，使用特点如下：

- 可以将导出接口改名，包括 default。
- 复合使用 export 与 import，也可以导出全部，当前模块导出的接口会覆盖继承导出的接口。

【示例 3-35】复合使用的示例如下。

```
export { foo, bar } from "methods";

// 约等于下面两段语句
import { foo, bar } from "methods";
export { foo, bar };

/* ------- 特点 1 --------*/
// 普通改名
export { foo as bar } from "methods";
// 将 foo 转导成 default
export { foo. as default } from "methods";
// 将 default 转导成 foo
export { default as foo } from "methods";

/* ------- 特点 2 --------*/
export * from "methods";
```

编写稍大一点的程序时一般都会将代码模块化。在 Node.js 中，一般将代码合理拆分到不同的 JavaScript 文件中，每一个文件就是一个模块，而文件路径就是模块名。

3.8 使用 Babel 转码

Babel 是一个广泛使用的转码器,可以将 ES6 代码转为 ES5 代码,从而在现有环境执行。这意味着可以使用 ES6 编写程序,而不用担心现有环境是否支持。例如,转码之前的原始代码用了箭头函数,这个特性还没有得到广泛支持,Babel 将其转为普通函数,就能在现有的 JavaScript 环境执行了:

```
// 转码前
input.map(item => item + 1);

// 转码后
input.map(function (item) {
  return item + 1;
});
```

再比如:

```
// 转码前
a=>a+1;

//转码后
(function (a) {
  return a + 1;
});
```

上述示例代码使用了箭头函数,Babel 将其转为普通函数。Babel 虽然支持浏览器环境,但网页实时将 ES6 代码转为 ES5 代码会对网页性能产生影响,所以我们需要使用构建工具在生产环境中将 ES6 代码进行提前转码。Babel 只能转换语法(如箭头函数),不支持新的全局变量,如 Iterator、Generator、Set、Maps、Proxy、Reflect、Symbol、Promise 等。如果想让这些方法运行,就必须使用 babel-polyfill 为当前环境提供该方法。

babel、babel-polyfill 安装:

```
$ yarn add babel-preset-env --dev
$ yarn add babel-polyfill --dev
```

安装完成后,我们可以通过.babelrc 文件或者 package.json 文件对 babel 进行配置。配置文件可以根据具体的需求进行配置,例如通过.babelrc 文件进行配置,新建.babelrc 文件并添加以下字段:

```
{
  "presets": [["env",{"useBuiltIns": true}]]
}
```

或者通过在 package.json 文件中增加字段:

```
{
  // ...
  "babel": {
    "presets": [["env",{"useBuiltIns": true}]]
  }
}
```

Babel 提供了 babel-cli 工具，可用于命令行转码，babel-cli 工具安装命令如下：

```
$ yarn add babel-cli --dev
```

babel-cli 基本用法：

```
# 输出转码结果
$ babel index.js

# 单文件转码
# --out-file 或 -o 参数指定输出文件
$ babel index.js --out-file compiled.js
$ babel index.js -o compiled.js

# 整个目录转码
# --out-dir 或 -d 参数指定输出目录
$ babel src --out-dir build
$ babel src -d build

# -s 参数生成 source map 文件
$ babel src -d build -s
```

我们可以在 package.json 文件中增加脚本：

```
{
  // ...
  "scripts": {
    "build": "babel src -d build"
  },
}
```

转码时执行如下命令：

```
$ yarn run build
```

babel-cli 工具自带一个 babel-node 命令，提供一个支持 ES6 的 REPL 环境，而且可以直接运行 ES6 代码。

执行 babel-node 就可以进入 REPL 环境：

```
$ babel-node
> (x => x * 2)(1)
2
```

babel-node 命令可以直接运行 ES6 脚本。将上面的代码放入脚本文件 test.js，然后直接运行：

```
$ babel-node test.js
2
```

我们可以改写 package.json：

```
{
  // ...
  "scripts": {
    "script-name": "babel-node test.js"
  }
}
```

用 babel-node 代替 node，test.js 就不需要做任何转码处理了。

3.9 使用 N-API

在 Node.js 开发领域中，原生 C++模块的开发一直是一个被人冷落的角落。但是实际上在必要的时候，用 C++进行 Node.js 的原生模块开发能有意想不到的好处：

- 性能提升。很多情况下，使用 C++进行 Node.js 原生模块开发性能会比纯 Node.js 开发要高，少数情况除外。
- 节约开发成本。在一些既有的 C++代码上做封装，开发成本远远低于从零开始编写 Node.js 代码。
- Node.js 无法完成的工作。个别情况下，开发者只能得到一个库的静态链接库或者动态链接库以及一堆 C++头文件，其余都是黑盒的，这种情况就不得不使用 C++进行模块开发了。

早期的 Node.js 原生 C++模块开发方式发生了多次变迁。自从 Node.js v8.0.0 发布之后，Node.js 推出了全新的用于开发 C++ 原生模块的接口（N-API），目的在于解决 Node.js 跨版本之间原生模块的兼容问题，NAN（Native Abstractions for Node.js）原生模块无法兼容不同版本的 Node.js，需要重新编译。N-API 做了一层抽象，实现了跨版本支持。v10.0 之前，N-API 是试验模式，使用该模块的时候会在 stderr（也可能是 stdout）输出一段警告，程序不依赖控制台可以无视此警告。v10.0 以后默认支持，警告不再出现。

本节对 Node.js 原生模块开发接口 N-API 做初步的尝试和解析，讲解 Node.js 原生 C++模块的开发。不同版本的 Node.js 使用同样的接口，这些接口是稳定地 ABI 化的，即应用二进制接口（Application Binary Interface）。这使得在不同版本的 Node.js 下，只要 ABI 的版本号一致，编译好的 C++扩展就可以直接使用，而不需要重新编译。事实上，在支持 N-API 接口的 Node.js

中，的确指定了当前 Node.js 所使用的 ABI 版本。

为了使得以后的 C++扩展开发、维护更方便，N-API 致力于以下的几个目标：

- 以 C 的风格提供稳定 ABI 接口。
- 消除 Node.js 版本的差异。
- 消除 JavaScript 引擎的差异（如 GoogleV8、MicrosoftChakraCore 等）。

为了达成上述隐藏的目标，N-API 的设计如下：

- 提供头文件 node_api.h。
- 任何 N-API 调用都返回一个 napi_status 枚举，来表示这次调用成功与否。
- N-API 的返回值为 napi_status，真实返回值由传入的参数来继承，如传入一个指针让函数操作。
- 所有 JavaScript 数据类型都被黑盒类型 napi_value 封装，不再是类似于 v8::Object、v8::Number 等类型。
- 如果函数调用不成功，就可以通过 napi_get_last_error_info 函数来获取最后一次出错的信息。

【示例 3-36】

（1）首先，生成一个项目目录：

```
n-api_test/
|-------- /test.js
|-------- /addon.c
|---------/binding.gyp
|---------/common.h
```

> **提 示**
>
> common.h 是一个包括插件使用的预定义宏的文件。

（2）然后开始编写 binding.gyp：

```
//binding.gyp
{
  "targets": [
    {
      "target_name": "addon",
      "sources": [ "addon.c" ]
    }
  ]
}
```

（3）编写 addon 插件：

```
//addon.c
```

```c
#include <node_api.h>
#include <string.h>
#include "./common.h"

napi_value Method(napi_env env, napi_callback_info info) {
  napi_value world;
  const char* str = "world";
  size_t str_len = strlen(str);
  NAPI_CALL(env, napi_create_string_utf8(env, str, str_len, &world));
  return world;
}

napi_value Init(napi_env env, napi_value exports) {
  napi_property_descriptor desc = DECLARE_NAPI_PROPERTY("hello", Method);
  NAPI_CALL(env, napi_define_properties(env, exports, 1, &desc));
  return exports;
}

NAPI_MODULE(NODE_GYP_MODULE_NAME, Init)


//common.h
#define NAPI_RETVAL_NOTHING  // Intentionally blank #define

#define GET_AND_THROW_LAST_ERROR(env)
  do {
    const napi_extended_error_info *error_info;
    napi_get_last_error_info((env), &error_info);
    bool is_pending;
    napi_is_exception_pending((env), &is_pending);
    /* If an exception is already pending, don't rethrow it */
    if (!is_pending) {
      const char* error_message = error_info->error_message != NULL ?
        error_info->error_message :
        "empty error message";
      napi_throw_error((env), NULL, error_message);
    }
  } while (0)

#define NAPI_ASSERT_BASE(env, assertion, message, ret_val)
  do {
    if (!(assertion)) {
      napi_throw_error(
          (env),
        NULL,
          "assertion (" #assertion ") failed: " message);
      return ret_val;
    }
  } while (0)

// Returns NULL on failed assertion.
// This is meant to be used inside napi_callback methods.
#define NAPI_ASSERT(env, assertion, message)
```

```
  NAPI_ASSERT_BASE(env, assertion, message, NULL)

// Returns empty on failed assertion.
// This is meant to be used inside functions with void return type.
#define NAPI_ASSERT_RETURN_VOID(env, assertion, message)                \
  NAPI_ASSERT_BASE(env, assertion, message, NAPI_RETVAL_NOTHING)

#define NAPI_CALL_BASE(env, the_call, ret_val)                          \
  do {                                                                  \
    if ((the_call) != napi_ok) {                                        \
      GET_AND_THROW_LAST_ERROR((env));                                  \
      return ret_val;                                                   \
    }                                                                   \
  } while (0)

// Returns NULL if the_call doesn't return napi_ok.
#define NAPI_CALL(env, the_call)                                        \
  NAPI_CALL_BASE(env, the_call, NULL)

// Returns empty if the_call doesn't return napi_ok.
#define NAPI_CALL_RETURN_VOID(env, the_call)                            \
  NAPI_CALL_BASE(env, the_call, NAPI_RETVAL_NOTHING)

#define DECLARE_NAPI_PROPERTY(name, func)                               \
  { (name), 0, (func), 0, 0, 0, napi_default, 0 }

#define DECLARE_NAPI_GETTER(name, func)                                 \
  { (name), 0, 0, (func), 0, 0, napi_default, 0 }
```

(4) 配置并编译文件：

```
$ node-gyp configure &&node-gyp build
```

(5) 然后编写测试的 JavaScript 文件：

```
//test.js
var assert = require('assert');
var addon = require(`./build/Release/addon.node`);
console.log(addon.hello());
```

(6) 运行测试文件：

```
$node test.js
```

获得正确的输出：

```
world
(node:27082) Warning: N-API is an experimental feature and could change at any time.
```

在最新版的 NAPI 设计中，甚至直接使用 C 作为使用语言，极大地降低了开发难度。

第 4 章 Node.js常用模块

本章介绍 Node.js 的模块机制以及常用模块，包括 Buffer、File System、HTTP 服务、Response 对象、TCP 服务、SSL 模块、WebSocket 模块、Stream 模块、Events 模块以及 RESTful API 等。

4.1 Module

为了让 Node.js 的文件可以相互调用，Node.js 提供了一个简单的模块系统。模块是 Node.js 应用程序的基本组成部分，文件和模块是一一对应的。换言之，一个 Node.js 文件就是一个模块，这个文件可能是 JavaScript 代码、JSON 或者编译过的 C/C++ 扩展。

4.1.1 创建和使用模块

在 Node.js 中，创建一个模块非常简单，我们创建一个 main.js 文件，代码如下：

```
var hello=require('./hello');
hello.world();
```

以上示例中，代码 require('./hello') 引入了当前目录下的 hello.js 文件（./为当前目录，node.js 默认后缀为 js）。

Node.js 提供了 exports 和 require 两个对象，其中 exports 是模块公开的接口，require 用于从外部获取一个模块的接口，即所获取模块的 exports 对象。接下来我们就来创建 hello.js 文件，代码如下：

```
exports.world=function(){
  console.log('HelloWorld');
}
```

在以上示例中，hello.js 通过 exports 对象把 world 作为模块的访问接口，在 main.js 中通过 require('./hello')加载这个模块，然后就可以直接访问 hello.js 中 exports 对象的成员函数了。

有时候我们只是想把一个对象封装到模块中，格式如下：

```
module.exports=function(){
//...
```

}
```

【示例 4-1】例如：

```
//hello.js
function Hello() {
   var name;
   this.setName = function(thyName) {
      name = thyName;
   };
   this.sayHello = function() {
      console.log('Hello ' + name);
   };
};
module.exports = Hello;
```

这样就可以直接获得这个对象了：

```
//main.js
var Hello = require('./hello');
hello = new Hello();
hello.setName('BYVoid');
hello.sayHello();
```

模块接口的唯一变化是使用 module.exports=Hello 代替了 exports.world=function(){}。在外部引用该模块时，其接口对象就是要输出的 Hello 对象本身，而不是原先的 exports。

在 Node.js 中引入模块，需要经历如下 3 个步骤：

- 路径分析。
- 文件定位。
- 编译执行。

在 Node.js 中，模块分为两类：一类是 Node.js 提供的模块，称为核心模块；另一类是用户编写的模块，称为文件模块。核心模块部分在 Node.js 源代码的编译过程中编译进了二进制执行文件。在 Node 进程启动时，部分核心模块就被直接加载进内存中，所以这部分核心模块引入时，文件定位和编译执行这两个步骤可以省略掉，并且在路径分析中优先判断，它的加载速度是最快的。文件模块则是在运行时动态加载，需要完整的路径分析、文件定位、编译执行过程，速度比核心模块慢。

4.1.2　require 方法中的文件查找策略

由于 Node.js 中存在 4 类模块（原生模块和 3 种文件模块），尽管 require 方法极其简单，但是内部的加载却是十分复杂的，其加载优先级也各自不同，如图 4.1 所示。
```

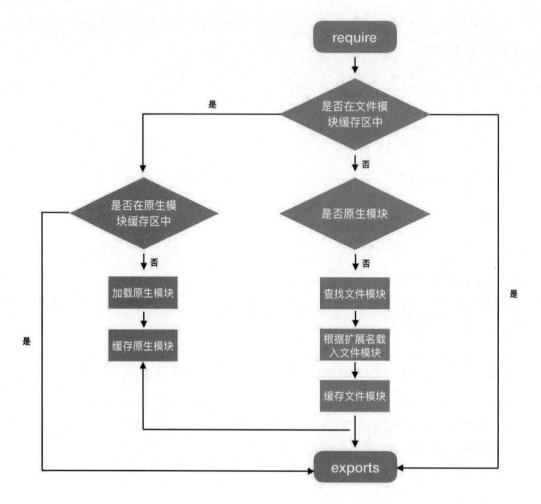

图 4.1　模块加载优先级

1. 从文件模块缓存中加载

尽管原生模块与文件模块的优先级不同,但是都会优先从文件模块的缓存中加载已经存在的模块。Node.js 对引入过的模块都会进行缓存,以减少二次引入时的开销。类似于在浏览器中,浏览器会缓存静态脚本文件以提高性能一样,Node.js 中不同的地方在于,浏览器仅仅缓存文件,而 Node.js 缓存的是编译和执行之后的对象。不论是核心模块还是文件模块,require()方法对相同模块的二次加载都一律采用缓存优先的方式,这是第一优先级的。不同之处在于核心模块的缓存检查先于文件模块的缓存检查。

2. 从原生模块加载

原生模块的优先级仅次于文件模块缓存的优先级。require 方法在解析文件名之后,优先检查模块是否在原生模块列表中。以 http 模块为例,尽管在目录下存在一个 http/http.js/http.node/http.json 文件,但是 require("http")不会从这些文件中加载,而是从原生模块中加载。

原生模块也有一个缓存区，同样也是优先从缓存区加载。如果缓存区没有被加载过，就调用原生模块的加载方式进行加载和执行。

3. 从文件加载

当文件模块缓存中不存在，而且不是原生模块的时候，Node.js 会解析 require 方法传入的参数，并从文件系统中加载实际的文件。require 方法接收以下几种参数的传递：

- http、fs、path 等，原生模块。
- ./mod 或 ../mod，相对路径的文件模块。
- /pathtomodule/mod，绝对路径的文件模块。
- mod，非原生模块的文件模块。

在进入路径查找之前，有必要描述一下 module path 这个 Node.js 中的概念。对于每一个被加载的文件模块，创建这个模块对象的时候，这个模块便会有一个 paths 属性，其值根据当前文件的路径计算得到。创建 modulepath.js 文件，其内容为：

```
console.log(module.paths);
```

我们将其放到任意一个目录中执行 node modulepath.js 命令，将得到以下的输出结果：

```
[ '/home/jackson/research/node_modules',
'/home/jackson/node_modules',
'/home/node_modules',
'/node_modules' ]
```

可以看出 module path 的生成规则为：从当前文件目录开始查找 node_modules 目录；然后依次进入父目录，查找父目录下的 node_modules 目录；依次迭代，直到根目录下的 node_modules 目录。

除此之外，还有一个全局 module path，是当前 node 执行文件的相对目录(../../lib/node)。如果在环境变量中设置了 HOME 目录和 NODE_PATH 目录，那么整个路径还包含 NODE_PATH 和 HOME 目录下的 .node_libraries 与 .node_modules。其最终值大致如下：

```
[NODE_PATH,HOME/.node_modules,HOME/.node_libraries,execPath/../../lib/node]
```

整个文件查找流程如图 4.2 所示。

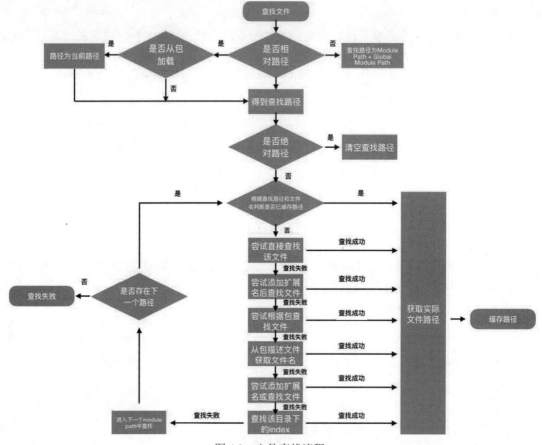

图 4.2　文件查找流程

如果 require 绝对路径的文件,查找时就不会去遍历每一个 node_modules 目录,其速度最快。其余流程如下:

(1)从 module path 数组中取出第一个目录作为查找基准。

(2)直接从目录中查找该文件,如果存在,就结束查找。如果不存在,就进行下一条查找。

(3)尝试添加.js、.json、.node 后缀后查找,如果存在文件,就结束查找。如果不存在,就进行下一条查找。

(4)尝试将 require 的参数作为一个包来进行查找,读取目录下的 package.json 文件,取得 main 参数指定的文件。

(5)尝试查找该文件,如果存在,就结束查找。如果不存在,就进行第 3 条查找。

(6)如果继续失败,就取出 module path 数组中的下一个目录作为基准查找,循环第 1~5 个步骤。

(7)如果继续失败,就循环第 1~6 个步骤,直到 module path 中的最后一个值。

(8)如果仍然失败,就抛出异常。

整个查找过程十分类似原型链的查找和作用域的查找。所幸 Node.js 对路径查找实现了缓存机制,否则由于每次判断路径都是同步阻塞式进行的,会导致严重的性能消耗。

4.2 Buffer

JavaScript 语言只有字符串数据类型,没有二进制数据类型。但是在 Node.js 中,应用需要处理图片、TCE 流、接收和上传文件、操作数据库,即在网络流和文件的操作过程中必须使用到二进制数据。JavaScript 自有的字符串类型无法满足大量二进制数据的处理,因此在 Node.js 中定义了 Buffer 类,该类用来创建一个专门存放二进制数据的缓存区。

在 Node.js 中,Buffer 类是随 Node 内核一起发布的核心库。Buffer 库为 Node.js 带来了一种存储原始数据的方法,可以让 Node.js 处理二进制数据,每当需要在 Node.js 中处理 I/O 操作中移动的数据时,就有可能使用 Buffer 库。Buffer 类似于一个整数数组,但它对应于 V8 堆内存之外的一块原始内存。Buffer 代表一个缓冲区,主要用于操作二进制数据流,其用法与数组非常相似。

4.2.1 Buffer 与字符编码及转换

Buffer 实例一般用于表示编码字符的序列,如 UTF-8、UCS2、Base64 或十六进制编码的数据。通过使用显式的字符编码就可以在 Buffer 实例与普通的 JavaScript 字符串之间进行相互转换,例如:

```
const buf = Buffer.from('runoob', 'ascii');

// 输出 72756e6f6f62
console.log(buf.toString('hex'));

// 输出 cnVub29i
console.log(buf.toString('base64'));
```

Node.js 目前支持的字符编码包括:

- ascii: 仅支持 7 位 ASCII 数据。如果设置去掉高位的话,那么这种编码是非常快的。
- utf8: 多字节编码的 Unicode 字符。许多网页和其他文档格式都使用 UTF-8。
- utf16le: 2 或 4 个字节,小端序编码的 Unicode 字符,支持代理对(U+10000~U+10FFFF)。
- ucs2: utf16le 的别名。
- base64: Base64 编码。
- latin1: 一种把 Buffer 编码成一字节编码的字符串的方式。
- binary: latin1 的别名。

- hex：将每个字节编码为两个十六进制字符。

4.2.2 Buffer 类及其方法

本节介绍 Buffer 类的创建及相关方法的使用。

1. 创建 Buffer 类

Buffer 存在于全局对象上，无须引入模块即可使用。Buffer 提供了如下方法来创建 Buffer 类：

- Buffer.alloc(size[, fill[, encoding]])：返回一个指定大小的 Buffer 实例，如果没有设置 fill，就默认填满 0。
- Buffer.allocUnsafe(size)：返回一个指定大小的 Buffer 实例，但是它不会被初始化，所以它可能包含敏感的数据。
- Buffer.allocUnsafeSlow(size)：创建一个大小为 size 字节的新 Buffer。
- Buffer.from(array)：返回一个被 array 的值初始化的新的 Buffer 实例（传入的 array 的元素只能是数字，不然就会自动被 0 覆盖）。
- Buffer.from(arrayBuffer[, byteOffset[, length]])：返回一个新建的与给定的 ArrayBuffer 共享同一内存的 Buffer。
- Buffer.from(buffer)：复制传入的 Buffer 实例的数据，并返回一个新的 Buffer 实例。
- Buffer.from(string[, encoding])：返回一个被 string 的值初始化的新的 Buffer 实例。

【示例 4-2】示例代码如下：

```
// 创建一个长度为10且用0填充的 Buffer
const buf1 = Buffer.alloc(10);

// 创建一个长度为10且用0×1 填充的Buffer
const buf2 = Buffer.alloc(10, 1);

// 创建一个长度为10且未初始化的 Buffer
// 这个方法比调用 Buffer.alloc() 更快
// 但返回的 Buffer 实例可能包含旧数据
// 因此需要使用 fill() 或 write() 重写
const buf3 = Buffer.allocUnsafe(10);

// 创建一个包含 [0x1, 0x2, 0x3] 的 Buffer
const buf4 = Buffer.from([1, 2, 3]);

// 创建一个包含 UTF-8 字节 [0x74, 0xc3, 0xa9, 0x73, 0x74] 的 Buffer
const buf5 = Buffer.from('tést');

// 创建一个包含 Latin-1 字节 [0x74, 0xe9, 0x73, 0x74] 的 Buffer
```

```
const buf6 = Buffer.from('tést', 'latin1');
```

一个 Buffer 对象的大小在创建时就固定下来了，创建之后不可改变，这个结论我们可以通过下面一个实例来说明：

```
var buf=new Buffer(5);
buf[6]=5;
console.log(buf)
```

运行以上代码，通过返回结果我们发现 buffer 对象的长度并没有改变：

```
<Buffer 00 00 00 00 00>
```

2. Buffer 写入

Buffer 写入的语法如下：

```
buf.write(string[, offset[, length]][, encoding])
```

即根据 encoding 的字符编码写入 string 到 buf 中的 offset 位置。length 参数是写入的字节数。如果 buf 没有足够的空间保存整个字符串，就只会写入 string 的一部分。只部分解码的字符不会被写入。该方法返回实际写入的大小。参数含义如下：

- String：写入的字符串。
- Offset：开始写入的索引值，默认为 0。
- Length：写入的字节数，默认为 buffer.length。
- Encoding：使用的编码，默认为'utf8'。

【示例 4-3】Buffer 写入的示例代码如下：

```
buf = Buffer.alloc(256);
len = buf.write("www.runoob.com");

console.log("写入字节数 : "+ len);
```

执行以上代码，输出结果为：

```
$node main.js
写入字节数 : 14
```

3. 从 Buffer 中读取数据

读取 Buffer 中的数据语法如下：

```
buf.toString([encoding[, start[, end]]])
```

从 Buffer 读取数据，该方法的返回值是解码缓冲区数据并使用指定的编码返回字符串。参数描述如下：

- Encoding：使用的编码，默认为 utf8。

- Start: 指定开始读取的索引位置，默认为 0。
- End: 结束位置，默认为 Buffer 的末尾。

【示例 4-4】读取数据的示例如下：

```
buf = Buffer.alloc(26);
for (var i = 0 ; i < 26 ; i++) {
  buf[i] = i + 97;
}

console.log( buf.toString('ascii'));          // 输出：abcdefghijklmnopqrstuvwxyz
console.log( buf.toString('ascii',0,5));      // 输出：abcde
console.log( buf.toString('utf8',0,5));       // 输出：abcde
console.log( buf.toString(undefined,0,5));    // 使用 UTF8 编码，并输出：abcde
```

执行以上代码，输出结果为：

```
$ node main.js
abcdefghijklmnopqrstuvwxyz
abcde
abcde
abcde
```

4. 将 Buffer 转换为 JSON 对象

将 Buffer 转换为 JSON 对象的函数语法格式如下：

```
buf.toJSON()
```

返回值是一个 JSON 对象。当字符串化一个 Buffer 实例时，JSON.stringify()会隐式地调用该 toJSON()。

【示例 4-5】例如如下代码：

```
const buf = Buffer.from([0x1, 0x2, 0x3, 0x4, 0x5]);
const json = JSON.stringify(buf);

// 输出: {"type":"Buffer","data":[1,2,3,4,5]}
console.log(json);

const copy = JSON.parse(json, (key, value) => {
  return value && value.type === 'Buffer' ?
    Buffer.from(value.data) :
    value;
});

// 输出: <Buffer 01 02 03 04 05>
console.log(copy);
```

执行以上代码，输出结果为：

```
{"type":"Buffer","data":[1,2,3,4,5]}
<Buffer 01 02 03 04 05>
```

5. Buffer 合并

Buffer 合并的语法如下：

```
Buffer.concat(list[, totalLength])
```

返回值是合并多个成员的一个新 Buffer 对象。参数说明：

- List：用于合并的 Buffer 对象数组列表。
- totalLength：指定合并后 Buffer 对象的总长度。

例如：

```
var buffer1 = Buffer.from(('Hello'));
var buffer2 = Buffer.from(('World'));
var buffer3 = Buffer.concat([buffer1,buffer2]);
console.log("buffer3 内容: " + buffer3.toString());
```

执行以上代码，输出结果为：

```
buffer3 内容: Hello World
```

6. Buffer 比较

Buffer 比较的函数语法如下：

```
buf.compare(otherBuffer);
```

该方法返回一个数字，表示 buf 在 otherBuffer 之前、之后或相同。其参数描述如下：

- otherBuffer：与 buf 对象比较的另一个 Buffer 对象。

【示例 4-6】示例代码如下：

```
var buffer1 = Buffer.from('ABC');
var buffer2 = Buffer.from('ABCD');
var result = buffer1.compare(buffer2);

if(result < 0) {
   console.log(buffer1 + " 在 " + buffer2 + "之前");
}else if(result == 0){
   console.log(buffer1 + " 与 " + buffer2 + "相同");
}else {
   console.log(buffer1 + " 在 " + buffer2 + "之后");
}
```

执行以上代码，输出结果为：

```
ABC 在 ABCD 之前
```

7. Buffer 拷贝

复制 Buffer 的语法如下：

```
buf.copy(targetBuffer[, targetStart[, sourceStart[, sourceEnd]]])
```

该方法没有返回值，参数描述如下：

- targetBuffer：要拷贝的 Buffer 对象。
- targetStart：数字，可选，默认为 0。
- sourceStart：数字，可选，默认为 0。
- sourceEnd：数字，可选，默认为 buffer.length。

【示例 4-7】示例代码如下：

```
var buf1 = Buffer.from('abcdefghijkl');
var buf2 = Buffer.from('RUNOOB');

//将 buf2 插入 buf1 指定位置上
buf2.copy(buf1, 2);

console.log(buf1.toString());
```

执行以上代码，输出结果为：

```
abRUNOOBijkl
```

8. Buffer 裁剪

Buffer 裁剪的语法如下：

```
buf.slice([start[, end]])
```

该方法返回一个新的缓冲区，它和旧缓冲区指向同一块内存，但是从索引 start 到 end 的位置剪切。参数描述如下：

- start：数字，可选，默认为 0。
- end：数字，可选，默认为 buffer.length。

【示例 4-8】示例代码如下：

```
var buffer1 = Buffer.from('runoob');
// 剪切缓冲区
var buffer2 = buffer1.slice(0,2);
console.log("buffer2 content: " + buffer2.toString());
```

执行以上代码，输出结果为：

```
buffer2 content: ru
```

9. Buffer 长度

Buffer 长度计算的语法如下：

```
buf.length;
```

返回 Buffer 对象所占据的内存长度。示例代码如下：

```
var buffer = Buffer.from('www.runoob.com');
// 缓冲区长度
console.log("buffer length: " + buffer.length);
```

执行以上代码，输出结果为：

```
buffer length: 14
```

4.2.3 Buffer 与性能

在 Node.js 中，进行 HTTP 传输时，若返回的类型为 string，则会将 string 类型的参数转换为 Buffer，通过 Node.js 中的 Stream 流一点点地返回给客户端。如果我们直接返回 Buffer 类型，就没有了转换操作，直接返回减少了 CPU 的重复使用率。

因此，我们在写业务代码时，部分资源可以预先转换为 Buffer 类型（如 JS、CSS 等静态资源文件），直接返回 buffer 给客户端，再比如一些文件转发的场景，将获取到的内容储存为 Buffer 直接转发，避免额外的转换操作。

4.3 File System

Node.js 内置的 fs 模块就是文件系统模块，负责提供对文件的操作，包括文件的读写等。Node.js 导入文件系统模块（fs 模块）的语法如下：

```
var fs = require("fs")
```

fs 模块提供的 API 基本上可以分为以下 3 类：

- 文件属性读写：其中常用的有 fs.stat、fs.chmod、fs.chown 等。
- 文件内容读写：其中常用的有 fs.readFile、fs.readdir、fs.writeFile、fs.mkdir 等。
- 底层文件操作：其中常用的有 fs.open、fs.read、fs.write、fs.close 等。

和所有其他 JavaScript 模块不同的是，fs 模块同时提供了异步和同步的方法。例如，读取文件内容的函数有异步的 fs.readFile()和同步的 fs.readFileSync()。异步的方法函数最后一个参数为回调函数，回调函数的第一个参数包含错误信息（error）。由于 Node 环境执行的 JavaScript 代码是服务器端代码，因此绝大部分需要在服务器运行期反复执行业务逻辑的代码，必须使用异步代码，否则同步代码在执行时期，因为 JavaScript 只有一个执行线程，服务器将停止响应。服务器启动时如果需要读取配置文件，或者结束时需要写入状态文件时，可以使用同步代码，因为这些代码只在启动和结束时执行一次，不影响服务器正常运行时的异步执行。

以下针对 fs 模块常用的 API 进行详细介绍。

4.3.1 异步读文件

【示例 4-9】首先创建 sample.txt 文件，内容如下：

```
Hello World!
文件读取实例
```

按照 JavaScript 的标准，异步读取一个文本文件的代码如下：

```
'use strict';

var fs = require('fs');

fs.readFile('sample.txt', 'utf-8', function (err, data) {
   if (err) {
      console.log(err);
   } else {
      console.log("异步读取: " + data.toString());
   }
});
```

注意，sample.txt 文件必须在当前目录下，且文件编码为 UTF-8。

异步读取时，传入的回调函数接收两个参数，当正常读取时，err 参数为 null，data 参数为读取到的 String。当读取发生错误时，err 参数代表一个错误对象，data 为 undefined。这也是 Node.js 标准的回调函数：第一个参数代表错误信息，第二个参数代表结果。后续我们还会经常编写这种回调函数。

由于 err 是否为 null 是判断是否出错的标志，因此通常的判断逻辑是：

```
if (err) {
   // 出错了
} else {
   // 正常
}
```

【示例 4-10】再比如，读取一个图片文件：

```
'use strict';
```

```
var fs = require('fs');

fs.readFile('sample.png', function (err, data) {
   if (err) {
      console.log(err);
   } else {
      console.log(data);
      console.log(data.length + ' bytes');
   }
});
```

当读取二进制文件时，如果不传入文件编码，回调函数的 data 参数就会返回一个 Buffer 对象。在 Node.js 中，Buffer 对象就是一个包含零个或任意个字节的数组（注意和 Array 不同）。

Buffer 对象可以和 String 进行转换，例如把一个 Buffer 对象转换成 String：

```
// Buffer -> String
var text = data.toString('utf-8');
console.log(text);
```

或者把一个 String 转换成 Buffer：

```
// String -> Buffer
var buf = Buffer.from(text, 'utf-8');
console.log(buf);
```

Buffer 的操作详见 4.2 节。

4.3.2 同步读文件

除了标准的异步读取模式外，fs 模块也提供相应的同步读取函数。同步读取的函数和异步函数相比多了一个 Sync 后缀，并且不接收回调函数，函数直接返回结果。

【示例 4-11】用 fs 模块同步读取一个文本文件的代码如下：

```
'use strict';

var fs = require('fs');

var data = fs.readFileSync('sample.txt', 'utf-8');
console.log(data);
```

可见，原异步调用的回调函数的 data 被函数直接返回，函数名需要改为 readFileSync，其他参数不变。

如果同步读取文件发生错误，就需要用 try...catch 捕获该错误：

```
try {
   var data = fs.readFileSync('sample.txt', 'utf-8');
```

```
      console.log(data);
} catch (err) {
   // 出错了
}
```

4.3.3 打开文件

异步模式下,打开文件的语法格式如下:

```
fs.open(path, flags[, mode], callback)
```

参数使用说明如下:

- path: 文件的路径。
- flags: 文件打开的行为。具体值详见下文。
- mode: 设置文件模式(权限),文件创建默认权限为 0666(可读,可写)。
- callback: 回调函数,带有两个参数,如 callback(err, fd)。

其中,flags 参数的取值如表 4.1 所示。

表 4.1 flags 参数的取值

方法	描述
r	以读取模式打开文件。如果文件不存在就抛出异常
r+	以读写模式打开文件。如果文件不存在就抛出异常
rs	以同步的方式读取文件
rs+	以同步的方式读取和写入文件
w	以写入模式打开文件,如果文件不存在就创建
wx	类似于'w',但是如果文件路径存在,文件就会写入失败
w+	以读写模式打开文件,如果文件不存在就创建
wx+	类似于'w+',但是如果文件路径存在,文件读写就失败
a	以追加模式打开文件,如果文件不存在就创建
ax	类似于'a',但是如果文件路径存在,文件追加就会失败
a+	以读取追加模式打开文件,如果文件不存在就创建
ax+	类似于'a+',但是如果文件路径存在,文件读取追加就会失败

【示例 4-12】打开 input.txt 文件进行读写的代码示例如下:

```
var fs = require("fs");

// 异步打开文件
console.log("准备打开文件!");
fs.open('input.txt', 'r+', function(err, fd) {
   if (err) {
      return console.error(err);
   }
```

```
    console.log("文件打开成功！");
});
```

以上代码执行结果如下：

```
$ node file.js
准备打开文件！
文件打开成功！
```

4.3.4 写入文件

将数据写入文件的异步方法是 fs.writeFile()。异步模式下，写入文件的语法格式如下：

```
fs.writeFile(file, data[, options], callback)
```

writeFile 直接打开文件默认是 w 模式，所以如果文件存在，该方法写入的内容就会覆盖旧的文件内容。

参数使用说明如下：

- file：文件名或文件描述符。
- data：要写入文件的数据，可以是 String（字符串）或 Buffer（缓冲）对象。
- options：该参数是一个对象，包含{encoding, mode, flag}。默认编码为 UTF-8，模式为 0666，flag 为'w'。
- callback：回调函数，回调函数只包含错误信息参数（err），在写入失败时返回。

【示例 4-13】示例代码如下：

```
'use strict';

var fs = require('fs');

var data = 'Hello, Node.js';
fs.writeFile('output.txt', data, function (err) {
    if (err) {
        console.log(err);
    } else {
        console.log('ok.');
    }
});
```

writeFile()的参数依次为文件名、数据和回调函数。如果传入的数据是 String 就默认按 UTF-8 编码写入文本文件，如果传入的参数是 Buffer，写入的就是二进制文件。由于回调函数只关心成功与否，因此只需要一个 err 参数。

和 readFile()类似，writeFile()也有一个同步方法，即 writeFileSync()：

```
'use strict';
```

```
var fs = require('fs');

var data = 'Hello, Node.js';
fs.writeFileSync('output.txt', data);
```

4.3.5 获取文件信息

如果需要获取文件大小、创建时间等信息，就可以使用 fs.stat()，该方法返回一个 Stat 对象，包含文件或目录的详细信息，语法格式如下：

```
fs.stat(path, callback)
```

参数使用说明如下：

- path：文件路径。
- callback：回调函数，带有两个参数，如(err, stats)，stats 是 fs.Stats 对象。

【示例 4-14】示例代码如下：

```
'use strict';

var fs = require('fs');

fs.stat('sample.txt', function (err, stat) {
    if (err) {
        console.log(err);
    } else {
        // 是否是文件
        console.log('isFile: ' + stat.isFile());
        // 是否是目录
        console.log('isDirectory: ' + stat.isDirectory());
        if (stat.isFile()) {
            // 文件大小
            console.log('size: ' + stat.size);
            // 创建时间, Date 对象
            console.log('birth time: ' + stat.birthtime);
            // 修改时间, Date 对象
            console.log('modified time: ' + stat.mtime);
        }
    }
});
```

fs.stat(path)执行后，会将 stat 类的实例返回给其回调函数。可以通过 stat 类中的提供方法判断文件的相关属性。该示例代码运行结果如下：

```
isFile: true
```

```
isDirectory: false
size: 181
birth time: Fri Dec 11 2015 09:43:41 GMT+0800 (CST)
modified time: Fri Dec 11 2015 12:09:00 GMT+0800 (CST)
```

stat 类中的方法详细见表 4.2 所示。

表 4.2　stat 方法说明

方法	描述
stat.isFile()	如果是文件就返回 true，否则返回 false
stat.isDirectory()	如果是目录就返回 true，否则返回 false
stat.isBlockDevice()	如果是块设备就返回 true，否则返回 false
stat.isCharacterDevice()	如果是字符设备就返回 true，否则返回 false
stat.isSymbolicLink()	如果是软链接就返回 true，否则返回 false
stat.isFIFO()	如果是 FIFO 就返回 true，否则返回 false。FIFO 是 UNIX 中的一种特殊类型的命令管道
stat.isSocket()	如果是 Socket 就返回 true，否则返回 false

stat() 也有一个对应的同步函数 statSync()，此处不再赘述。

4.3.6　fs.read 异步读文件

在异步模式下，读取文件的语法格式如下：

```
fs.read(fd, buffer, offset, length, position, callback)
```

该方法使用了文件描述符来读取文件，其参数含义如下：

- fd：通过 fs.open() 方法返回的文件描述符。
- buffer：数据写入的缓冲区。
- offset：缓冲区写入的写入偏移量。
- length：要从文件中读取的字节数。
- position：文件读取的起始位置，如果 position 的值为 null，就会从当前文件指针的位置读取。
- callback：回调函数，有 3 个参数：err、bytesRead 和 buffer，err 为错误信息，bytesRead 表示读取的字节数，buffer 为缓冲区对象。

【示例 4-15】例如，input.txt 文件内容为：

```
Hello World!
```

读取文件的示例代码如下：

```
var fs = require("fs");
var buf = new Buffer.alloc(1024);
```

```
console.log("准备打开已存在的文件！");
fs.open('input.txt', 'r+', function(err, fd) {
   if (err) {
      return console.error(err);
   }
   console.log("文件打开成功！");
   console.log("准备读取文件：");
   fs.read(fd, buf, 0, buf.length, 0, function(err, bytes){
      if (err){
         console.log(err);
      }
      console.log(bytes + " 字节被读取");

      // 仅输出读取的字节
      if(bytes > 0){
         console.log(buf.slice(0, bytes).toString());
      }
   });
});
```

以上代码执行结果如下：

```
$ node file.js
准备打开已存在的文件！
文件打开成功！
准备读取文件：
42 字节被读取
Hello World!
```

4.3.7 fs.close 异步关闭文件

在异步模式下，关闭文件的语法格式如下：

```
fs.close(fd, callback)
```

该方法使用了文件描述符来读取文件，其参数使用说明如下：

- fd：通过 fs.open() 方法返回的文件描述符。
- callback：回调函数，没有参数。

【示例 4-16】关闭文件的代码如下：

```
var fs = require("fs");
var buf = new Buffer.alloc(1024);

console.log("准备打开文件！");
fs.open('input.txt', 'r+', function(err, fd) {
   if (err) {
```

```
      return console.error(err);
   }
   console.log("文件打开成功！");
   console.log("准备读取文件！");
   fs.read(fd, buf, 0, buf.length, 0, function(err, bytes){
      if (err){
         console.log(err);
      }

      // 仅输出读取的字节
      if(bytes > 0){
         console.log(buf.slice(0, bytes).toString());
      }

      // 关闭文件
      fs.close(fd, function(err){
         if (err){
            console.log(err);
         }
         console.log("文件关闭成功");
      });
   });
});
```

以上代码执行结果如下：

```
$ node file.js
准备打开文件！
文件打开成功！
准备读取文件！
Hello World!
文件关闭成功
```

4.4 HTTP/HTTP2 服务

浏览器和服务器之间的传输协议是 HTTP（Hyper Text Transfer Protocol），HTTP 是在网络上传输 HTML 的协议，用于浏览器和服务器之间的通信。要开发 HTTP 服务器程序，从头处理 TCP 连接，解析 HTTP 是不现实的。这些工作实际上已经由 Node.js 自带的 http 模块完成了。

应用程序并不直接和 HTTP 协议打交道，而是操作 http 模块提供的 request 和 response 对象。request 对象封装了 HTTP 请求，我们调用 request 对象的属性和方法就可以拿到所有 HTTP 请求的信息；response 对象封装了 HTTP 响应，我们操作 response 对象的方法就可以把 HTTP 响应返回给浏览器。

4.4.1　http 模块

http 模块提供两种使用方式：作为服务端使用时，创建一个 HTTP 服务器，监听 HTTP 客户端请求并返回响应；作为客户端使用时，发起一个 HTTP 客户端请求，获取服务端响应。

1. http 模块作为服务端使用

【示例 4-17】用 Node.js 实现一个简单的 HTTP 服务器，示例代码如下：

```
'use strict';

// 导入http模块
var http = require('http');

// 创建http server，并传入回调函数
var server = http.createServer(function (request, response) {
    // 回调函数接收 request 和 response 对象
    // 获得 HTTP 请求的 method 和 url
console.log(request.method + ': ' + request.url);
request.on('data', function (chunk) {
  body.push(chunk);
});
    // 将 HTTP 响应200写入 response，同时设置 Content-Type: text/html
    response.writeHead(200, {'Content-Type': 'text/html'});
    // 将 HTTP 响应的 HTML 内容写入 response
    response.end('<h1>Hello world!</h1>');
});

// 让服务器监听8080端口
server.listen(8080);

console.log('Server is running at http://127.0.0.1:8080/');
```

首先需要使用 .createServer 方法创建一个服务器，然后调用 .listen 方法监听端口。之后，每当来了一个客户端请求，创建服务器时传入的回调函数就被调用一次。再运行该程序，可以看到以下输出：

```
$ node 4.4.js
Server is running at http://127.0.0.1:8080/
```

直接打开浏览器输入 http://127.0.0.1:8080/，即可看到服务器响应的内容，如图 4.3 所示。

图 4.3　HTTP 服务器

同时，在命令提示符窗口可以看到程序打印的请求信息，如图 4.4 所示。

```
GET: /
GET: /favicon.ico
```

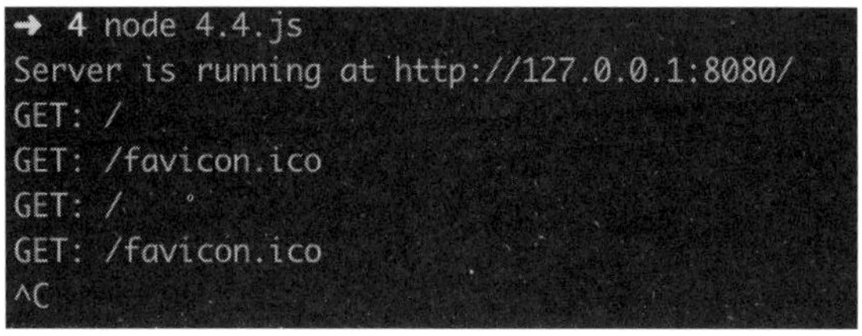

图 4.4　HTTP 响应

HTTP 请求本质上是一个数据流，由请求头（Headers）和请求体（Body）组成。例如以下是一个完整的 HTTP 请求数据内容。

```
// 请求头
POST / HTTP/1.1
User-Agent: curl/7.26.0
Host: localhost
Accept: */*
Content-Length: 11
Content-Type: application/x-www-form-urlencoded
// 请求体
Hello World
```

HTTP 请求在发送给服务器时，可以认为是按照从头到尾的顺序一个字节一个字节地以数据流方式发送的。而 http 模块创建的 HTTP 服务器在接收到完整的请求头后，就会调用回调函数。在回调函数中，除了可以使用 request 对象访问请求头数据外，还能把 request 对象当作一个只读数据流来访问请求体数据。

【示例 4-18】示例代码如下：

```
http.createServer(function (request, response) {
    var body = [];

    console.log(request.method);
    console.log(request.headers);

    request.on('data', function (chunk) {
        body.push(chunk);
    });

    request.on('end', function () {
        body = Buffer.concat(body);
        console.log(body.toString());
    });
```

```
}).listen(8080);
```

在回调函数中，除了可以使用 response 对象来写入响应头数据外，还能把 response 对象当作一个只写数据流来写入响应体数据。例如，在服务端原样将客户端请求的请求体数据返回给客户端：

```
http.createServer(function (request, response) {
    response.writeHead(200, { 'Content-Type': 'text/plain' });

    request.on('data', function (chunk) {
        response.write(chunk);
    });

    request.on('end', function () {
        response.end();
    });
}).listen(80);
```

http 模块在客户端模式下发起一个客户端 HTTP 请求，需要指定目标服务器的位置并发送请求头和请求体，例如：

```
var options = {
    hostname: 'www.example.com',
    port: 80,
    path: '/upload',
    method: 'POST',
    headers: {
        'Content-Type': 'application/x-www-form-urlencoded'
    }
};

var request = http.request(options, function (response) {});

request.write('Hello World');
request.end();
```

request 方法创建了一个客户端，并指定请求目标和请求头数据。之后，就可以把 request 对象当作一个只写数据流来写入请求体数据并结束请求。另外，由于在 HTTP 请求中，GET 请求是很常见的一种，并且不需要请求体，因此 http 模块也提供了以下便捷 API：

```
http.get('http://www.example.com/', function (response) {});
```

2. http 模块作为客户端使用

当客户端发送请求并接收到完整的服务端响应头时，就会调用回调函数。在回调函数中，除了可以使用 response 对象访问响应头数据外，还能把 response 对象当作一个只读数据流来访问响应体数据，例如：

```javascript
http.get('http://www.example.com/', function (response) {
   var body = [];

   console.log(response.statusCode);
   console.log(response.headers);

   response.on('data', function (chunk) {
      body.push(chunk);
   });

   response.on('end', function () {
      body = Buffer.concat(body);
      console.log(body.toString());
   });
});
```

4.4.2 http2 模块

HTTP/2 是新一代 HTTP 协议，支持多路复用（Multiplexing）、首部压缩（Headers）、服务端推送（Server Push）等特性，能够有效减少时延。

服务端推送即允许服务器在客户端缓存中填充数据，通过服务器推送的机制来提前请求。和使用 http 1.1 不同的是，首先需要安装 http2 模块：

```
npm i --save http2
```

【示例 4-19】使用 http2 模块的代码示例：

```javascript
'use strict'

const fs = require('fs')
const path = require('path')
// eslint-disable-next-line
const http2 = require('http2')
const helper = require('./helper')

const PORT = process.env.PORT || 8080
const PUBLIC_PATH = path.join(__dirname, '../public')

const publicFiles = helper.getFiles(PUBLIC_PATH)

//创建 HTTP2服务器
const server = http2.createSecureServer({
 cert: fs.readFileSync(path.join(__dirname, '../ssl/cert.pem')),
 key: fs.readFileSync(path.join(__dirname, '../ssl/key.pem'))
}, onRequest)

// Request 事件
function onRequest (req, res) {
   // 路径指向 index.html
```

```
  const reqPath = req.url === '/' ? '/index.html' : req.url
    //获取html资源
  const file = publicFiles.get(reqPath)

  // 文件不存在
  if (!file) {
    res.statusCode = 404
    res.end()
    return
  }

  res.stream.respondWithFD(file.fileDescriptor, file.headers)
}

server.listen(PORT, (err) => {
  console.log('监听服务器启动=====\n')
  if (err) {
    console.error(err)
    return
  }

  console.log(`Server listening on ${PORT}`)
})
```

引入 http2，使用 http2.createSecureServer()创建一个服务器，语法格式如下：

```
http2.createSecureServer(options, callback)
```

options 表示证书或者其他有关的配置选项，证书是必需的。

启动服务，然后交给浏览器渲染 HTML 和加载资源：

```
<!DOCTYPE html>
<html>

<head>
  <meta charset="UTF-8">
</head>

<body>
  <h1>体验 HTTP2</h1>
</body>
<script src="bundle1.js"></script>
<script src="bundle2.js"></script>
<script>
  for (var i = 0; i < 100; i++) {
    fetch('//localhost:8080/network.png');
  }
</script>

</html>
```

效果如图 4.5 所示。

第 4 章 Node.js 常用模块

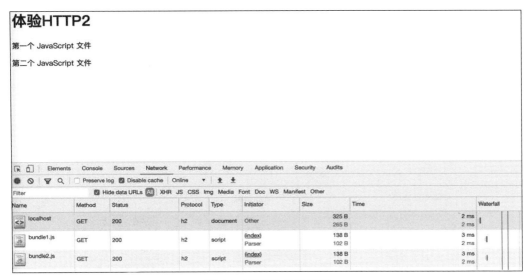

图 4.5　HTTP2 效果示例

不使用服务器推送与使用服务器推送的数据交互对比如图 4.6 所示。

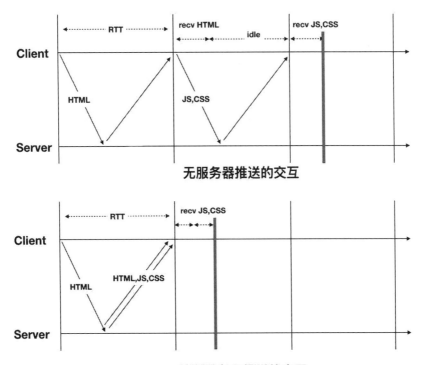

图 4.6　有/无服务器推送的交互对比

113

4.5 TCP 服务

互联网基于分层架构实现，包括应用层、传输层、网络层、链路层、物理层。图 4.7 显示了 OSI 七层模型与 TCP/IP 五层模型之间的关系，中间使用虚线标注了传输层，对于上层应用层（HTTP/HTTPS 等）也都是基于这一层的 TCP 协议来实现的，所以想使用 Node.js 做服务端开发，net 模块是必须掌握的。

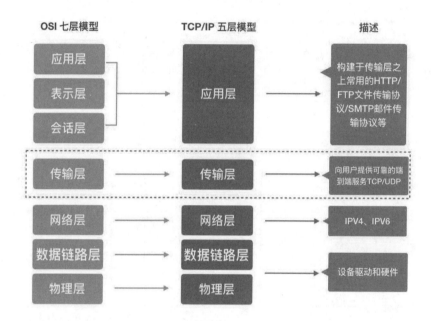

图 4.7　OSI 七层模型 与 TCP/IP 五层模型之间的关系

网络是互联网通信的基础，对于网络的每个层次，Node.js 基本都有对应的模块，如 https、http、net（TCP）、tls/crypto 等，其中 net、http、dgram 模块分别用来实现 TCP、HTTP、UDP 的通信。http 为应用层模块，主要按照特定协议编解码数据；net 为传输层模块，主要负责传输编码后的应用层数据；https 是一个综合模块（涵盖了 http/tls/crypto 等），主要用于确保数据安全性。本节介绍 TCP 服务器和 TCP 客户端的构建。net 模块是 Node.js 的核心模块。在 http 模块里可以看到 http.Server 继承了 net.Server，此外，HTTP 客户端与 HTTP 服务端的通信均依赖于 socket（net.Socket）。也就是说，做 Node 服务端编程，net 基本是绕不开的一个模块。

从组成来看，net 模块主要包含两部分：

- net.Server：TCP Server，内部通过 socket 来实现与客户端的通信。
- net.Socket：TCP/本地 socket 的 Node 版实现，它实现了全双工的 stream 接口。

4.5.1 构建 TCP 服务器

为了使用 Node.js 创建 TCP 服务器，首先要使用 require("net")来加载 net 模块，然后使用 net 模块的 createServer 方法就可以轻松地创建一个 TCP 服务器。

```
/**
 * 构建 TCP 客户端
 */

/* 引入 net 模块 */
var net = require("net");

/* 创建 TCP 服务器 */
var server = net.createServer(function(socket){
   console.log('someone connects');
})

/* 设置连接的服务器 */
server.listen(8000, function(){
   console.log("Creat server on http://127.0.0.1:8000/");
})
```

运行这段代码并使用浏览器访问 http://127.0.0.1:8000/，会看到控制台打印了"someone connects"，表明已经成功连接到这个 TCP 服务器。

```
/* 设置连接的服务器 */
server.listen(8000, function(){
console.log("Creat server on http://127.0.0.1:8000/");
})
```

上面这段代码实际上触发的是 server 下的 listening 事件，等同于：

```
/* 设置监听端口 */

server.listen(8000);

/* 设置监听时的回调函数 */

server.on("listening", function () {

 console.log("Creat server on http://127.0.0.1:8000/");

})
```

除了 listening 事件外，TCP 服务器还支持以下事件：

- listening：调用 server.listen()，当正式开始监听请求的时候触发。

- connection：当有新的连接创建时触发，回调函数的参数为 socket 连接对象。
- close：当 TCP 服务器关闭的时候触发，回调函数没有参数。
- error：当 TCP 服务器发生错误的时候触发，回调函数的参数为 error 对象，比如监听了已经被占用的端口。

【示例 4-20】以下代码通过 net.Server 类来创建一个 TCP 服务器，并添加以上事件。

```
/**
 * 通过 net.Server 类来创建一个 TCP 服务器
 */

/* 引入 net 模块 */
var net = require("net");

/* 实例化一个服务器对象 */
var server = new net.Server();

/* 监听 connection 事件 */
server.on("connection", function (socket) {
   console.log("someone connects");
});

/* 设置监听端口 */
server.listen(8000);

/* 设置监听时的回调函数 */
server.on("listening", function () {
   console.log("Creat server on http://127.0.0.1:8000/");
})

/* 设置关闭时的回调函数 */
server.on("close", function () {
   console.log("server closed!");
})

/* 设置错误时的回调函数 */
server.on("error", function (err) {
   console.log("error!");
})
```

当创建了一个 TCP 服务器后，可以通过 server.address() 方法来查看这个 TCP 服务器监听的地址，并返回一个 JSON 对象。这个对象的属性有：

- port：TCP 服务器监听的端口号。
- family：说明 TCP 服务器监听的地址是 IPv6 还是 IPv4。

- address:TCP 服务器监听的地址。

【示例 4-21】示例代码如下:

```
/**
 * 查看服务器监听的地址
 */

/* 引入 net 模块 */
var net = require("net");

/* 创建 TCP 服务器 */
var server = net.createServer(function(socket){
    console.log('someone connects');
})

/* 获取地址信息 */
server.listen(8000,function(){
    /* 获取地址信息,得到的是一个json { address: '::', family: 'IPv6', port: 8000 }
*/
    var address = server.address();

    /* TCP 服务器监听的端口号 */
    console.log("the port of server is"+ address.port);

    /* TCP 服务器监听的地址 */
    console.log("the address of server is"+ address.address);

    /* 说明 TCP 服务器监听的地址是 IPv6 还是 IPv4 */
    console.log("the family of server is"+ address.family);
})
```

创建一个 TCP 服务器后,可以通过 server.getConnections()方法获取连接这个 TCP 服务器的客户端数量。除此之外,也可以通过 maxConnections 属性来设置这个服务器的最大连接数量,当连接数量超过最大值时,服务器将拒绝新的连接,例如:

```
/**
 * 连接服务器的客户端数量
 */

/* 引入 net 模块 */
var net = require("net");

/* 创建 TCP 服务器 */
var server = net.createServer(function(socket){
    console.log('someone connects');
```

```
    /* 设置最大连接数量 */
    server.maxConnections=3;
    server.getConnections(function(err,count){
        console.log("the count of client is "+count);
    })
})

/* 获取监听端口 */
server.listen(8000,function(){
    console.log("Creat server on http://127.0.0.1:8000/");
})
```

4.5.2 服务器和客户端之间的通信

socket 对象可以用来获取客户端发出的流数据,每次接收到数据的时候触发 data 事件,通过监听这个事件就可以在回调函数中获取客户端发送的数据。

【示例 4-22】

```
/**
 * 发送和获取
 */

/* 引入 net 模块 */
var net = require("net");

/* 创建 TCP 服务器 */
var server = net.createServer(function(socket){
    /* 获取地址信息 */
    var address = server.address();
    var message = "the server address is"+JSON.stringify(address);

    /* 发送数据 */
    socket.write(message,function(){
        var writeSize = socket.bytesWritten;
        console.log(message + "has send");
        console.log("the size of message is"+writeSize);
    })

    /* 监听 data 事件 */
    socket.on('data',function(data){
        console.log(data.toString());
        var readSize = socket.bytesRead;
        console.log("the size of data is"+readSize);
    })
```

```
})
/* 获取地址信息 */
server.listen(8000,function(){
    console.log("Creat server on http://127.0.0.1:8000/");
})
```

TCP 服务器给客户端发送了字符串：

```
the server address is{"address":"::","family":"IPv6","port":8000}has send
```

客户端给 TCP 服务器发送了字符串 "hello TCP!" 和字节数，如图 4.8 和图 4.9 所示。

图 4.8

图 4.9

4.5.3 构建 TCP 客户端

可以用 Node.js 来构建一个 TCP 客户端，实现 TCP 客户端和 TCP 服务器的通信。

【示例 4-23】为了使用 Node.js 创建 TCP 客户端，首先要使用 require("net") 来加载 net 模块，然后创建一个连接 TCP 客户端的 socket 对象即可：

```
/* 引入 net 模块 */

var net = require("net");

/* 创建 TCP 客户端 */
```

```
var client = net.Socket();
```

创建完 socket 对象后，使用 socket 对象的 connect 方法就可以连接一个 TCP 服务器。

```
/**
 * 构建 TCP 客户端
 */

/* 引入 net 模块 */
var net = require("net");

/* 创建 TCP 客户端 */
var client = net.Socket();

/* 设置连接的服务器 */
client.connect(8000, '127.0.0.1', function () {
    console.log("connect the server");

    /* 向服务器发送数据 */
    client.write("message from client");
})

/* 监听服务器传来的 data 数据 */
client.on("data", function (data) {
    console.log("the data of server is " + data.toString());
})

/* 监听 end 事件 */
client.on("end", function () {
    console.log("data end");
})
```

运行服务器和客户端之间通信的示例代码，之后再运行 TCP 客户端示例代码，可以发现服务器已经接收到客户端的数据，客户端也已经接收到服务端的数据，如图 4.10 和图 4.11 所示。

```
➜  4 node 4-5/server5.js
Creat server on http://127.0.0.1:8000/
the server address is{"address":"::","family":"IPv6","port":8000}has send
the size of message is65
message from client
the size of data is19
the server address is{"address":"::","family":"IPv6","port":8000}has send
the size of message is65
```

图 4.10　TCP 通信之 Server 端

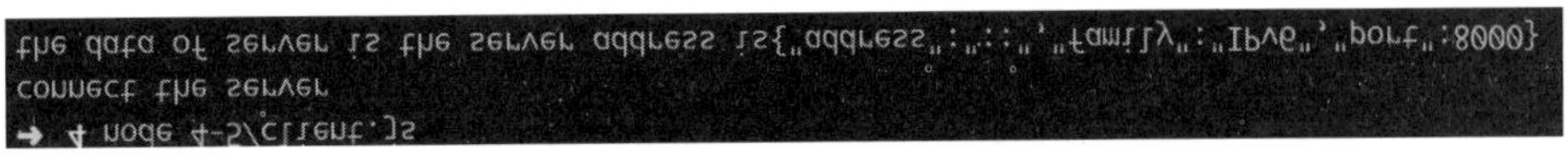

图 4.11 TCP 通信之 Client 端

net.Socket 连接相关的 API 有：

- socket.connect()：有 3 种不同的参数，用于不同的场景。
- socket.setTimeout()：用来进行连接超时设置。
- socket.setKeepAlive()：用来设置长连接。
- socket.destroy( )、socket.destroyed：当错误发生时，用来销毁 socket，确保这个 socket 上不会再有其他的 IO 操作。

net.Socket 涉及的事件：

- data：当收到另一侧传来的数据时触发。
- connect：当连接建立时触发。
- close：当连接断开时触发。如果是因为传输错误导致的连接断开，参数就为 error。
- end：当连接另一侧发送了 FIN 包的时候触发（读者可以回顾一下 HTTP 如何断开连接的）。默认情况下（allowHalfOpen == false），socket 会完成自我销毁操作。但也可以把 allowHalfOpen 设置为 true，这样就可以继续往 socket 里写数据。当然，最后需要手动调用 socket.end()。
- error：当有错误发生时就会触发，参数为 error。
- timeout：提示用户 socket 已经超时，需要手动关闭连接。
- drain：当写缓存空了的时候触发。
- lookup：当域名解析完成时触发。

4.6 SSL

HTTP（Hyper Text Transfer Protocol，超文本传输协议）是互联网上使用广泛的一种协议，是所有 WWW 文件必须遵循的标准。HTTP 协议传输的数据都是未加密的，也就是明文的，因此使用 HTTP 协议传输隐私信息非常不安全。HTTPS（Hyper Text Transfer Protocol over Secure Socket Layer，安全的超文本传输协议）使用 SSL（Secure Sockets Layer，安全套接层）协议用于对 HTTP 协议传输的数据进行加密，保证会话过程中的安全性。

4.6.1 SSL 简介

SSL 协议位于 TCP/IP 协议与各种应用层协议（如 HTTP）之间。SSL 协议可分为两层：
SSL Record Protocol（SSL 记录协议）建立在可靠的传输协议（如 TCP）上，为高层协议提供数据封装、压缩、加密等基本功能的支持。

SSL Handshake Protocol(SSL 握手协议)建立在记录协议上，用于在实际数据传输开始前，通信双方进行身份验证、协商加密算法、交换加密密钥等。

SSL 协议既用到了对称加密又用到了非对称加密（公钥加密），在建立传输链路时，SSL 首先对对称加密的密钥使用公钥进行非对称加密，链路建立好之后，SSL 对传输内容使用对称加密。对称加密和公钥加密对比如下：

（1）对称加密：速度高，可加密内容多，用来加密会话过程中的消息。

（2）公钥加密：加密速度慢，但能提供更好的身份认证技术，用来加密对称加密的密钥。

HTTPS 在建立 Socket 连接之前，需要进行握手，单向认证的具体过程参见表 4.3。

表 4.3　单向认证

客　户　端	单向认证	服　务　端
>>>	1. 发送客户端 SSL 版本等信息	>>>
<<<	2. 服务端给客户端返回 SSL 版本、随机数等信息，以及服务器公钥	<<<
CHECK	3. 客户端校验服务端证书是否合法，合法继续，否则警告	WAIT
>>>	4. 客户端发送自己可支持的对称加密方案给服务端供选择	>>>
WAIT	5. 服务端选择加密程度高的加密方式	CHECK
<<<	6. 将选择好的加密方案以明文方式发送给客户端	<<<
>>>	7. 客户端收到加密方式产生随机码，作为对称加密密钥，使用服务器公钥加密后发给服务器	>>>
WAIT	8. 服务端使用私钥对加密信息进行解密，获得对称加密密钥	CHECK
---	9. 使用对称加密进行通信确保安全	---

HTTPS 双向认证和单向认证的原理基本差不多，只是除了客户端需要认证服务端以外，增加了服务端对客户端的认证，具体过程参见表 4.4。

表 4.4　双向认证

客　户　端	双向认证	服　务　端
>>>	1. 发送客户端 SSL 版本等信息	>>>
<<<	2. 服务端给客户端返回 SSL 版本、随机数等信息，以及服务器公钥	<<<
CHECK	3. 客户端校验服务端证书是否合法，合法就继续，否则警告	WAIT
>>>	4. 客户端校验通过后，将自己的证书和公钥发送至客户端	>>>
WAIT	5. 对客户端校验，校验结束后获得客户端公钥	CHECK
>>>	6. 客户端发送自己可支持的对称加密方案给服务端供选择	>>>
WAIT	7. 服务端选择加密程度高的加密方式	CHECK
<<<	8. 将选择好的加密方案以客户端公钥进行加密后发送给客户端	<<<
>>>	9. 客户端收到加密方式，使用私钥进行解密，产生随机码，作为对称加密密钥，使用服务器公钥加密后发给服务器	>>>
WAIT	10. 服务端使用私钥对加密信息进行解密，获得对称加密密钥	CHECK
---	11. 使用对称加密进行通信确保安全	---

4.6.2 使用 OpenSSL 进行证书生成

由于 TLS（SSL）是基于非对称的加密体系，因此在开发前需要准备用于加密解密和验证的私钥、数字证书。这里分别为 CA、服务器、客户端准备一套密钥及证书。

1. 生成 CA 证书及密钥

（1）生成一个密钥为 ca-key.pem：

```
openssl genrsa -out ca-key.pem -des 1024
```

（2）根据密钥 ca-key.pem 生成一个证书签名请求文件 ca-csr.pem，根据这个文件可以签出密钥对应的证书。在这里会要求填入相关信息，可以用"."表示空，但 Common Name 必须填入对应域名，否则浏览器检查证书与签名域名不相符会不允许通过。

```
openssl req -new -key ca-key.pem -out ca-csr.pem
```

不执行第一步，直接执行以下语句，也会自动生成一个 private.pem 以及 ca-csr.pem。

```
openssl req -new -out ca-csr.pem
```

（3）利用证书签名请求文件 ca-csr.pem，并利用 ca-key.pem 签名生成证书 ca-cert.pem。其中 x509 为证书格式 X.509，-days 为有效期天数。

```
openssl x509 -req -days 3650 -in ca-csr.pem -signkey ca-key.pem -out ca-cert.pem
```

也可以拿着 ca-csr.pem 去权威的第三方申请一个证书，这样申请得到的证书是被接受的。现在我们自签自申证书，则如以上代码所示，使用证书签名请求文件签一个证书。

2. 生成服务端证书及密钥

（1）生成一个服务器私钥为 server-key.pem：

```
openssl genrsa -out server-key.pem 1024
```

（2）生成证书签名请求文件 server-csr.pem：

```
openssl req -new -key server-key.pem -out server-csr.pem
```

假如我们需要多 DNS 和多 IP 的主机使用同一个证书完成验证，就不能通过上面的命令来实现。我们需要一个 openssl.cnf 配置文件，并执行以下语句：

```
openssl req -new -key server-key.pem -config openssl.cnf -out server-csr.pem
```

openssl.cnf 配置文件内容格式如下：

```
[req]
    distinguished_name = req_distinguished_name
    req_extensions = v3_req

    [req_distinguished_name]
```

```
countryName = Country Name (2 letter code)
countryName_default = CN
stateOrProvinceName = State or Province Name (full name)
stateOrProvinceName_default = BeiJing
localityName = Locality Name (eg, city)
localityName_default = YaYunCun
organizationalUnitName = Organizational Unit Name (eg, section)
organizationalUnitName_default = Domain Control Validated
commonName = Internet Widgits Ltd
commonName_max = 64

[ v3_req ]
# Extensions to add to a certificate request
basicConstraints = CA:FALSE
keyUsage = nonRepudiation, digitalSignature, keyEncipherment
subjectAltName = @alt_names

[alt_names]
DNS.1 = ns1.dns.com
DNS.2 = ns2.dns.com
DNS.3 = ns3.dns.com
IP.1 = 192.168.1.84
IP.2 = 127.0.0.1
IP.3 = 127.0.0.2
```

（3）生成服务器证书为 server-sert.pem，更多选项含义可以在 OpenSSL 官网查询：

```
openssl x509 -req -days 730 -CA ca-cert.pem -CAkey ca-key.pem -CAcreateserial -in server-csr.pem -out server-cert.pem -extensions v3_req -extfile openssl.cnf
```

3. 生成客户端证书及密钥

（1）生成客户端密钥 client-key.pem：

```
openssl genrsa -out client-key.pem
```

（2）生成证书签名请求文件 client-csr.pem：

```
openssl req -new -key client-key.pem -out client-csr.pem
```

（3）生成服务器证书为 server-sert.pem，更多选项含义可以在 OpenSSL 官网查询：

```
openssl x509 -req -days 365 -CA ca-cert.pem -CAkey ca-key.pem -CAcreateserial -in client-csr.pem -out client-cert.pem
```

4. 打包和转换

（1）可以将服务器的私钥、证书、CA 证书打包成一个单独的 .pfx 或 .p12 文件以便于使用，如 .p12 导入浏览器可以让浏览器信任该证书。

```
    openssl pkcs12 -export -in server-cert.pem -inkey server-key.pem -certfile 
ca-cert.pem -out server.pfx
    openssl pkcs12 -export -in client-cert.pem -inkey client-key.pem -certfile 
ca-cert.pem -out client.p12
```

（2）证书的转换，由在上一步打包好的文件，可以经过转换变为其他格式，以下提供常见的格式转换方法：

```
openssl pkcs12 -export -inkey test.key -in test.cer -out test.pfx
openssl pkcs12 -in test.pfx -nodes -out test.pem
openssl rsa -in test.pem -out test.key
openssl x509 -in test.pem -out test.crt
```

4.6.3　Node.js 实现 HTTPS 的配置

（1）导入 https 和 fs：

```
var https = require ('https');        // https 服务器
const fs = require("fs");             // 文件输入输出，用来导入证书
```

（2）导入证书，利用 fs 导出 3 个证书并存入变量：

```
var privateKey = fs.readFileSync('./cert/ca.key').toString();
var certificate = fs.readFileSync('./cert/ca.crt').toString();
var client_certificate = fs.readFileSync('./cert/client.crt').toString();
```

（3）https 设置，利用键值对来进行设置，requestCert 表示是否需要客户端证书，rejectUnauthorized 表示如果客户端证书不符合是否拒绝访问，这两个为 true 则意味着双向认证，否则为单向认证：

```
var credentials = { key: privateKey,
                    cert: certificate,
                    ca: client_certificate,
                    requestCert: true,
                    rejectUnauthorized: true};
```

（4）创建服务器对象，比 http 多了一个参数，即第 3 步中所提的 credentials：

```
var server = http.createServer(app);                       // http 只需要一个参数就行
var server = https.createServer(credentials, app);  // https 需要的参数
```

（5）开启端口监听：

```
server.listen(port);                                       // port 即为端口号
```

4.7 WebSocket

WebSocket 是 HTML 5 开始提供的一种浏览器与服务器间进行全双工通信的网络技术。在 WebSocket API 中，浏览器和服务器只需要一次握手，之后浏览器和服务器之间就形成了一条快速通道，浏览器和服务器之间直接就可以互相传送数据了。

WebSocket 是一个通信协议，分为服务器和客户端。服务器放在后台，保持与客户端的长连接，完成双方通信的任务。客户端一般都是实现在支持 HTML 5 浏览器的核心中，通过提供 JavaScript API 使用网页可以建立 WebSocket 连接。

4.7.1 ws 模块

要使用 WebSocket，关键在于服务器端的支持，这样我们才有可能用支持 WebSocket 的浏览器使用 WebSocket。

【示例 4-24】

（1）在 Node.js 中，ws 是使用很广泛的 WebSocket 模块，首先在 package.json 中添加 ws 的依赖：

```
"dependencies": {
    "ws": "7.1.2"
}
```

（2）整个工程结构如下：

```
ws/
+- app.js <-- 启动 JS 文件
|
+- package.json <-- 项目描述文件
|
+- node_modules/ <-- npm 安装的所有依赖包
```

运行 npm install 后，我们就可以在 app.js 中编写 WebSocket 的服务器端代码。

（3）创建一个 WebSocket 的服务器实例：

```
// 导入 WebSocket 模块
const WebSocket = require('ws');

// 引用 Server 类
const WebSocketServer = WebSocket.Server;

// 实例化
const wss = new WebSocketServer({
```

```
    port: 3000
});
```

在 3000 端口上打开了一个 WebSocket Server，该实例由变量 wss 引用。

（4）接下来，如果有 WebSocket 请求接入，wss 对象可以响应 connection 事件来处理这个 WebSocket：

```
wss.on('connection', function (ws) {
    console.log('[SERVER] connection()');
    ws.on('message', function (message) {
        console.log('[SERVER] Received: ${message}');
        ws.send('ECHO: ${message}', (err) => {
            if (err) {
                console.log(`[SERVER] error: ${err}`);
            }
        });
    })
});
```

在 connection 事件中，回调函数会传入一个 WebSocket 的实例，表示这个 WebSocket 连接。对于每个 WebSocket 连接，我们都要对它绑定某些事件方法来处理不同的事件。这里，我们通过响应 message 事件，在收到消息后再返回一个 ECHO: xxx 的消息给客户端。

（5）接下来创建 WebSocket 连接。在命令行执行 npm start。在当前页面下，直接打开可以执行 JavaScript 代码的浏览器 Console，依次输入代码：

```
// 打开一个 WebSocket
var ws = new WebSocket('ws://localhost:3000/test');
// 响应 onmessage 事件
ws.onmessage = function(msg) { console.log(msg); };
// 给服务器发送一个字符串
ws.send('Hello!');
```

可以看到 Console 的输出如下：

```
MessageEvent {isTrusted: true, data: "ECHO: Hello!", origin: "ws://localhost:3000", lastEventId: "", source: null…}
```

这样，就在浏览器中成功地接收到了服务器发送的消息。如果觉得在浏览器中输入 JavaScript 代码比较麻烦，我们还可以直接用 ws 模块提供的 WebSocket 来充当客户端。换句话说，ws 模块既包含服务器端，又包含客户端。

（6）ws 的 WebSocket 就表示客户端，它其实是 WebSocketServer 响应 connection 事件时回调函数传入的变量 ws 的类型。

客户端的写法如下：

```
let ws = new WebSocket('ws://localhost:3000/test');

// 打开WebSocket连接后立刻发送一条消息
ws.on('open', function () {
    console.log(`[CLIENT] open()`);
    ws.send('Hello!');
});

// 响应收到的消息:
ws.on('message', function (message){
    console.log('[CLIENT] Received:${message}');
}
```

在 Node 环境下，ws 模块的客户端可以用于测试服务器端代码，否则，每次都必须在浏览器上执行 JavaScript 代码。

从上面的测试可以看出，WebSocket 协议本身不要求同源策略（Same-Origin Policy），也就是某个地址为 http://a.com 的网页可以通过 WebSocket 连接到 ws://b.com。但是，浏览器会发送 Origin 的 HTTP 头给服务器，服务器可以根据 Origin 拒绝这个 WebSocket 请求。所以，是否要求同源要看服务器端如何检查。

4.7.2 实战：ws 简易聊天室

本节使用 ws 模块创建一个简易聊天室，步骤如下：

（1）配置 package.json

```
{
  "name": "chat-server",
  "version": "0.0.1",
  "description": "my chat server",
  "dependencies": {
    "express": "^4.17.1",
    "http": "^0.0.0",
    "socket.io": "^2.2.0"
  }
}
```

（2）使用 npm 命令安装依赖

```
npm install
```

npm 包安装完后，工作目录下生成了一个名为 node_modules 的文件夹，里面分别是 express 和 socket.io，可能还有一些相关的依赖包。

（3）创建服务端

index.js 是支持 ws 通信的代码（服务端代码），监听 3000 端口（此端口在客户端脚本要使用，可以用 ip:3000 访问）；app.js 是搭建客户端访问网页的代码，监听 8080 端口：

```js
var app = require('express')();
var http = require('http').Server(app);
var io = require('socket.io')(http);

app.get('/', function(req, res){
    res.send('<h1>Welcome Realtime Server</h1>');
});

//在线用户
var onlineUsers = {};
//当前在线人数
var onlineCount = 0;

io.on('connection', function(socket){
    console.log('a user connected');

    //监听新用户加入
    socket.on('login', function(obj){
        //将新加入用户的唯一标识当作socket的名称，后面退出的时候会用到
        socket.name = obj.userid;

        //检查在线列表，如果不在里面就加入
        if(!onlineUsers.hasOwnProperty(obj.userid)) {
            onlineUsers[obj.userid] = obj.username;
            //在线人数加1
            onlineCount++;
        }

        //向所有客户端广播用户加入
        io.emit('login', {onlineUsers:onlineUsers, onlineCount:onlineCount, user:obj});
        console.log(obj.username+'加入了聊天室');
    });

    //监听用户退出
    socket.on('disconnect', function(){
        //将退出的用户从在线列表中删除
        if(onlineUsers.hasOwnProperty(socket.name)) {
            //退出用户的信息
            var obj = {userid:socket.name, username:onlineUsers[socket.name]};

            //删除
            delete onlineUsers[socket.name];
            //在线人数减1
```

```
            onlineCount--;

            //向所有客户端广播用户退出
            io.emit('logout',                    {onlineUsers:onlineUsers,
onlineCount:onlineCount, user:obj});
            console.log(obj.username+'退出了聊天室');
        }
    });

    //监听用户发布聊天内容
    socket.on('message', function(obj){
        //向所有客户端广播发布的消息
        io.emit('message', obj);
        console.log(obj.username+'说: '+obj.content);
    });

});

http.listen(3000, function(){
    console.log('listening on *:3000');
});
app.js 代码:
const express = require('express')
const path = require('path')
const app = express()
//使用静态资源访问,public 为根目录
app.use(express.static(path.join(__dirname, 'public')))

app.listen(8080, () => {
  console.log(`App listening at port 8080`)
})
```

（4）创建客户端

客户端有 3 个文件，index.html 是聊天室的 HTML 静态页面，client.js 是客户端连接 ws 通信的脚本，socket.io.js 是支持 socket.io 客户端的代码。

HTML 文件：

```
<!DOCTYPE html>
<html>
<head>
    <meta charset="utf-8">
    <meta name="format-detection" content="telephone=no"/>
    <meta name="format-detection" content="email=no"/>
    <meta content="width=device-width, initial-scale=1.0, maximum-scale=1.0, minimum-scale=1.0, user-scalable=0" name="viewport">
```

```html
    <title>T</title>
    <script src="https://cdn.bootcss.com/socket.io/2.0.4/socket.io.js"></script>
  </head>
  <body>
    <div id="loginbox">
        <div style="width:260px;margin:200px auto;">
            请先输入你在聊天室的昵称
            <br/>
            <br/>
            <input type="text" style="width:180px;" placeholder="请输入用户名" id="username" name="username" />
            <input type="button" style="width:50px;" value="提 交 " onclick="CHAT.usernameSubmit();"/>
        </div>
    </div>
    <div id="chatbox" style="display:none;">
        <div style="background:#3d3d3d;height: 28px; width: 100%;font-size:12px;">
            <div style="line-height: 28px;color:#fff;">
                <span style="text-align:left;margin-left:10px;">Websocket 多人聊天室</span>
                <span style="float:right; margin-right:10px;"><span id="showusername"></span> |
                    <a href="javascript:;" onclick="CHAT.logout()" style="color:#fff;">退出</a></span>
            </div>
        </div>
        <div id="doc">
            <div id="chat">
                <div id="message" class="message">
                    <div id="onlinecount" style="background:#EFEFF4; font-size:12px; margin-top:10px; margin-left:10px; color:#666;">
                    </div>
                </div>
                <div class="input-box">
                    <div class="input">
                        <input type="text" maxlength="140" placeholder="请输入聊天内容，按 Ctrl 提交" id="content" name="content">
                    </div>
                    <div class="action">
                        <button type="button" id="mjr_send" onclick="CHAT.submit();">提交</button>
                    </div>
                </div>
            </div>
```

```
        </div>
    </div>
    <script type="text/javascript" src="./client.js"></script>
</body>
</html>
```

客户端代码:

```
(function () {
    var d = document,
        w = window,
        p = parseInt,
        dd = d.documentElement,
        db = d.body,
        dc = d.compatMode == 'CSS1Compat',
        dx = dc ? dd: db,
        ec = encodeURIComponent;


    w.CHAT = {
        msgObj:d.getElementById("message"),
        screenheight:w.innerHeight ? w.innerHeight : dx.clientHeight,
        username:null,
        userid:null,
        socket:null,
        //让浏览器滚动条保持在最低部
        scrollToBottom:function(){
            w.scrollTo(0, this.msgObj.clientHeight);
        },
        //退出，本例只是一个简单的刷新
        logout:function(){
            //this.socket.disconnect();
            location.reload();
        },
        //提交聊天消息内容
        submit:function(){
            var content = d.getElementById("content").value;
            if(content != ''){
                var obj = {
                    userid: this.userid,
                    username: this.username,
                    content: content
                };
                this.socket.emit('message', obj);
                d.getElementById("content").value = '';
            }
```

```javascript
        return false;
    },
    genUid:function(){
        return new Date().getTime()+""+Math.floor(Math.random()*899+100);
    },
    //更新系统消息，本例中在用户加入、退出的时候调用
    updateSysMsg:function(o, action){
        //当前在线用户列表
        var onlineUsers = o.onlineUsers;
        //当前在线人数
        var onlineCount = o.onlineCount;
        //新加入用户的信息
        var user = o.user;

        //更新在线人数
        var userhtml = '';
        var separator = '';
        for(key in onlineUsers) {
            if(onlineUsers.hasOwnProperty(key)){
                userhtml += separator+onlineUsers[key];
                separator = '、';
            }
        }
        d.getElementById("onlinecount").innerHTML = '当前共有 '+onlineCount+' 人在线，在线列表：'+userhtml;

        //添加系统消息
        var html = '';
        html += '<div class="msg-system">';
        html += user.username;
        html += (action == 'login') ? ' 加入了聊天室' : ' 退出了聊天室';
        html += '</div>';
        var section = d.createElement('section');
        section.className = 'system J-mjrlinkWrap J-cutMsg';
        section.innerHTML = html;
        this.msgObj.appendChild(section);
        this.scrollToBottom();
    },
    //第一个界面用户提交用户名
    usernameSubmit:function(){
        var username = d.getElementById("username").value;
        if(username != ""){
            d.getElementById("username").value = '';
            d.getElementById("loginbox").style.display = 'none';
            d.getElementById("chatbox").style.display = 'block';
```

```
            this.init(username);
        }
        return false;
    },
    init:function(username){
        /*
        客户端根据时间和随机数生成uid,这样使得聊天室用户名称可以重复
        在实际项目中,如果需要用户登录,那么直接采用用户的uid来做标识就可以
        */
        this.userid = this.genUid();
        this.username = username;

        d.getElementById("showusername").innerHTML = this.username;
        this.msgObj.style.minHeight = (this.screenheight - db.clientHeight
+ this.msgObj.clientHeight) + "px";
        this.scrollToBottom();

        //连接WebSocket后端服务器
        this.socket = io.connect('ws://139.199.163.196:3000/');

        //告诉服务器端有用户登录
        this.socket.emit('login',                         {userid:this.userid,
username:this.username});

        //监听新用户登录
        this.socket.on('login', function(o){
            CHAT.updateSysMsg(o, 'login');
        });

        //监听用户退出
        this.socket.on('logout', function(o){
            CHAT.updateSysMsg(o, 'logout');
        });

        //监听消息发送
        this.socket.on('message', function(obj){
            var isme = (obj.userid == CHAT.userid) ? true : false;
            var contentDiv = '<div>'+obj.content+'</div>';
            var usernameDiv = '<span>'+obj.username+'</span>';

            var section = d.createElement('section');
            if(isme){
                section.className = 'user';
                section.innerHTML = contentDiv + usernameDiv;
            } else {
```

```
                section.className = 'service';
                section.innerHTML = usernameDiv + contentDiv;
            }
            CHAT.msgObj.appendChild(section);
            CHAT.scrollToBottom();
        });


    }
};
//通过回车键提交用户名
d.getElementById("username").onkeydown = function(e) {
    e = e || event;
    if (e.keyCode === 13) {
        CHAT.usernameSubmit();
    }
};
//通过回车键提交信息
d.getElementById("content").onkeydown = function(e) {
    e = e || event;
    if (e.keyCode === 13) {
        CHAT.submit();
    }
};
})();
```

(5) 运行程序

至此，所有编码已经完成，在项目目录执行：

```
node index.js
node app.js
```

执行后服务端和客户端效果如图 4.12 和图 4.13 所示。

图 4.12　聊天室控制台输出

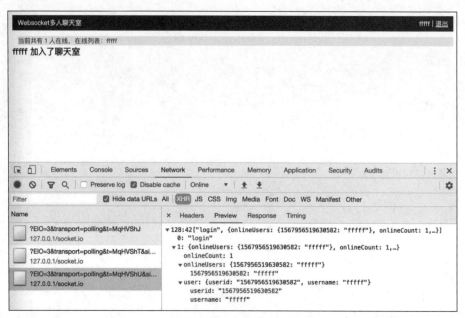

图 4.13　WebSocket 示例

4.8　流

流（Stream）是一种数据传输手段，是有顺序的，它有起点和终点。有时无须关心文件的主体内容，只关心能不能取到数据，取到数据之后怎么进行处理。对于小型的文本文件，我们可以把文件内容全部读入内存，再写入文件。对于体积较大的二进制文件，如音频、视频文件，以 GB 为单位来衡量大小，如果使用这种方法，会很容易使内存"爆仓"。针对大文件，理想的方法应该是读一部分，写一部分，无论文件有多大，只要时间允许，总会处理完成，这里就需要用到流的概念。

在 Node.js 中，流是一个抽象接口，Node.js 中有很多对象的实现都是流，如 HTTP 服务器 request 和 response 对象等。

Node.js 中有 4 种基本的流类型：

- Readable：可读的流，例如 fs.createReadStream()。
- Writable：可写的流，例如 fs.createWriteStream()。
- Duplex：可读写的流，例如 net.Socket。
- Transform：在读写过程中可以修改和变换数据的 Duplex 流。

可以通过 require('stream') 加载 Stream 基类。其中包括 Readable 流、Writable 流、Duplex 流和 Transform 流的基类。

4.8.1 可读流

可读流（Readable Stream）是对提供数据的源头（Source）的抽象，可读流的实现有：

- HTTP responses, on the client：客户端请求。
- HTTP requests, on the server：服务端请求。
- fs read streams：读文件。
- zlib streams：压缩。
- crypto streams：加密。
- TCP sockets：TCP 协议。
- child process stdout and stderr：子进程标准输出和错误输出。
- process.stdin：标准输入。

【示例 4-25】所有的 Readable 都实现了 stream.Readable 类定义的接口。示例代码如下：

```
let fs = require('fs');
//通过创建一个可读流
let rs = fs.createReadStream('./1.txt',{
    flags:'r',               //我们要对文件进行何种操作
    mode:0o666,              //权限位
    encoding:'utf8',         //不传默认为buffer,显示为字符串
    start:3,                 //从索引为3的位置开始读
    //这是我见过的唯一一个包括结束索引的
    end:8,                   //读到索引为8结束
    highWaterMark:3          //缓冲区大小
});
rs.on('open',function () {
    console.log('文件打开');
});
rs.setEncoding('utf8');      //显示为字符串
//希望流有一个暂停和恢复触发的机制
rs.on('data',function (data) {
    console.log(data);
    rs.pause();              //暂停读取和发射data事件
    setTimeout(function(){
        rs.resume();         //恢复读取并触发data事件
    },2000);
});
//如果读取文件出错了,就会触发error事件
rs.on('error',function () {
    console.log("error");
});
//如果文件的内容读完了,就会触发end事件
rs.on('end',function () {
    console.log('读完了');
```

```
});
rs.on('close',function () {
    console.log('文件关闭');
});
```

运行以上代码,输出:

```
文件打开
334
455
读完了
文件关闭
```

在代码中,监听可读流的 data 事件,一旦开始监听 data 事件,流就可以读文件的内容并且传出 data,边读边传输。默认情况下,当监听 data 事件之后,会不停地读数据、触发 data 事件,触发 data 事件后再次读数据。读的时候不是把文件整体内容读出来再传输出来的,而且设置一个缓冲区,大小默认是 64KB,若文件是 128KB,先读 64KB 传输出来,再读 64KB 传输出来,会传输两次。该缓冲区的大小可以通过 highWaterMark 来设置。

4.8.2 可写流

可写流(Writable Stream)是对数据写入目的地的一种抽象。可写流的例子包括:

- HTTP requests, on the client: 客户端请求。
- HTTP responses, on the server: 服务器响应。
- fs write streams: 文件。
- zlib streams: 压缩。
- crypto streams: 加密。
- TCP sockets TCP: 服务器。
- child process stdin: 子进程标准输入。
- process.stdout, process.stderr: 标准输出,错误输出。

【示例 4-26】所有可写流都实现了 stream.Writable 类定义的接口。尽管可写流的具体实例可能略有差别,但所有的可写流都遵循同一基本的使用模式,示例如下:

```
let fs = require('fs');
let ws = fs.createWriteStream('./2.txt',{
    flags:'w',
    mode:0o666,
    start:3,
    highWaterMark:3         //默认是16KB
});
```

当往可写流里写数据的时候,是不会立刻写入文件的,而是会先写入缓存区,缓存区的大小就是 highWaterMark，默认值是 16KB，等缓存区满了之后再次真正地写入文件里。如果缓存区已满，就返回 false，如果缓存区未满，就返回 true。如果能接着写，就返回 true，如果不能接着写，就返回 false。

```
let flag = ws.write('1');
console.log(flag);//true
flag =ws.write('2');
console.log(flag);//true
flag =ws.write('3');
console.log(flag);//false
flag =ws.write('4');
console.log(flag);//false
```

如果调用 stream.write(chunk)方法返回 false，流就会在适当的时机触发 drain 事件，这时才可以继续向流中写入数据，例如：

```
let fs = require('fs');
let ws = fs.createWriteStream('2.txt',{
    flags:'w',
    mode:0o666,
    start:0,
    highWaterMark:3
});
let count = 9;
function write(){
 let flag = true;//缓存区未满
    //写入方法是同步的,但是写入文件的过程是异步的。在真正写入文件后,还会执行我们的回调函数
 while(flag && count>0){
    console.log('before',count);
    flag = ws.write((count)+'','utf8',(function (i) {
        return ()=>console.log('after',i);
    })(count));
    count--;
 }
}
write();//987
//监听缓存区清空事件
ws.on('drain',function () {
    console.log('drain');
    write();//654 321
});
ws.on('error',function (err) {
    console.log(err);
});
/**
```

```
before 9
before 8
before 7
after 9
after 8
after 7
**/
```

如果已经不再需要写入了，就可以调用 end 方法关闭写入流，一旦调用 end 方法之后，就不能再写入。若在"ws.end();"后写"ws.write('x');"，则会报错"write after end"。

4.8.3 管道流

管道提供了一个输出流到输入流的机制。通常我们用于从一个流中获取数据并将数据传递到另一个流中。

【示例 4-27】例如，我们通过读取一个文件内容并将内容写入另一个文件中。

（1）首先创建 input.txt 文件：

```
Hello World!
OK!
```

（2）读取 input.txt 文件中的内容并写入 output.txt 文件中：

```
var fs = require("fs");

// 创建一个可读流
var readerStream = fs.createReadStream('input.txt');

// 创建一个可写流
var writerStream = fs.createWriteStream('output.txt');

// 管道读写操作
// 读取 input.txt 文件内容，并将内容写入 output.txt 文件中
readerStream.pipe(writerStream);

console.log("end");
```

代码执行结果如下：

```
$ node main.js
程序执行完毕
```

（3）查看 output.txt 文件的内容：

```
$ cat output.txt
Hello World!
```

OK!

（4）使用管道和链式来压缩和解压文件：

```
var fs = require("fs");
var zlib = require('zlib');

// 压缩 input.txt 文件为 input.txt.gz
fs.createReadStream('input.txt')
  .pipe(zlib.createGzip())
  .pipe(fs.createWriteStream('input.txt.gz'));

console.log("文件压缩完成。");
```

代码执行结果如下：

```
$ node compress.js
文件压缩完成。
```

执行完以上操作后，我们可以看到当前目录下生成了 input.txt 的压缩文件 input.txt.gz。

（5）接下来，让我们来解压该文件，创建 decompress.js 文件，代码如下：

```
var fs = require("fs");
var zlib = require('zlib');

// 解压 input.txt.gz 文件为 input.txt
fs.createReadStream('input.txt.gz')
  .pipe(zlib.createGunzip())
  .pipe(fs.createWriteStream('input.txt'));

console.log("文件解压完成。");
```

代码执行结果如下：

```
$ node decompress.js
文件解压完成。
```

4.9 事件

Node.js 主要 API 使用异步事件驱动模型，异步 I/O 操作完成时，某些类型对象（触发器）会周期性地触发一个命名事件到事件队列，用来调用函数对象（监听器）。例如，net.Server 对象在每次有新连接时触发一个事件，fs.readStream 对象在文件打开时触发一个事件。所有这些能产生事件的对象都是 events.EventEmitter 的实例。events 模块只提供了一个对象：events.EventEmitter。EventEmitter 的核心就是事件触发与事件监听器功能的封装。

通过 require("events")引入 events 模块：

```
// 引入 events 模块
var events = require('events');
// 创建 eventEmitter 对象
var eventEmitter = new events.EventEmitter();
```

若 EventEmitter 对象在实例化时发生错误，则会触发 error 事件。当添加新的监听器时，newListener 事件会触发，当监听器被移除时，removeListener 事件被触发。

4.9.1 注册事件名&监听器

通过 on()注册一个事件（名）和一个监听器，添加到监听器数组末尾，addListener()是 on()的别名。emit()按顺序执行每个监听器。

EventEmitter 的用法示例：

```
//event.js 文件
var EventEmitter = require('events').EventEmitter;
var event = new EventEmitter();
event.on('some_event', function() {
   console.log('some_event 事件触发');
});
setTimeout(function() {
   event.emit('some_event');
}, 1000);
```

event 对象注册了事件 some_event 的一个监听器，通过 setTimeout 在 1000 毫秒以后向 event 对象发送事件 some_event，此时会调用 some_event 的监听器。运行这段代码，1 秒后控制台输出了"some_event 事件触发"。

EventEmitter 的每个事件由一个事件名和若干个参数组成，事件名是一个字符串，通常表达一定的语义。对于每个事件，EventEmitter 支持多个事件监听器。当事件触发时，注册到这个事件的事件监听器被依次调用，事件参数作为回调函数参数传递。

【示例 4-28】多个时间监听器的示例：

```
//event.js 文件
var events = require('events');
var emitter = new events.EventEmitter();
emitter.on('someEvent', function(arg1, arg2) {
   console.log('listener1', arg1, arg2);
});
emitter.on('someEvent', function(arg1, arg2) {
   console.log('listener2', arg1, arg2);
});
emitter.emit('someEvent', 'arg1 参数', 'arg2 参数');
```

上述代码中，emitter 为事件 someEvent 注册了两个事件监听器，触发了 someEvent 事件。运行的结果如下：

```
$ node event.js
listener1 arg1 参数 arg2 参数
listener2 arg1 参数 arg2 参数
```

4.9.2 给监听器 listener 传入参数与 this

【示例 4-29】emit 方法允许给监听器传入任意参数，并且标准 this 会被设置为监听器所附加的 EventEmitter。

```
var events = require('events');
var eventEmitter = new events.EventEmitter;
eventEmitter.on('start', function(a, b) {
   console.log('eventEmitter starts');
   console.log(a, b, this);
});
eventEmitter.emit('start', 'a', 'b');
```

运行结果如下：

```
eventEmitter starts
a b { domain: null,
  _events: { start: [Function] },
  _maxListeners: 10 }
```

4.9.3 最多只触发一次的监听器

【示例 4-30】on 注册的监听器可以触发多次，once 注册的最多触发一次。

```
var events = require('events');
var eventEmitter = new events.EventEmitter;
//用 on 注册事件，绑定监听器
eventEmitter.on('start', function() {
   console.log('eventEmitter starts');
});
eventEmitter.emit('start');
eventEmitter.emit('start');

//用 once 注册事件，绑定监听器
eventEmitter.once('do', function() {
   console.log('eventEmitter do');
});
eventEmitter.emit('do');
eventEmitter.emit('do');
```

运行结果如下：

```
eventEmitter starts
eventEmitter starts
eventEmitter do
```

4.9.4 添加监听器/移除监听器事件

【示例4-31】当新增监听器时，所有 EventEmitter 会触发 newListener 事件；当移除监听器时，会触发 removeListener 事件。

```
var events = require('events');
var eventEmitter = new events.EventEmitter;
//用 once 注册事件, 只会执行一次
eventEmitter.once('newListener', function(event, listener) {
   if(event=='start') {
       eventEmitter.on('start',function(){
          console.log('start 2');
       });
       console.log('start');
   }
});
eventEmitter.on('start',function(){
   console.log('eventEmitter start');
});

eventEmitter.emit('start');
```

运行输出：

```
start
start 2
eventEmitter start
```

可以看到，触发 start 事件会同时触发 newListener 事件，并且 on 注册事件不会等待，而是直接执行后面的语句。

4.9.5 错误事件

EventEmitter 实例发生错误时，会触发名为 error 的特殊事件。它包含错误的语义，在遇到异常的时候通常会触发 error 事件。当 error 事件被触发时，EventEmitter 规定如果没有响应的监听器，Node.js 会把它当作异常，退出程序并输出错误信息。

【示例4-32】为防止 node.js 程序崩溃，建议始终为 error 事件注册监听器，例如：

```
var events = require('events');
var eventEmitter = new events.EventEmitter;
```

```
//用 on 注册事件，绑定监听器
eventEmitter.on('error', function(err) {
   console.log('eventEmitter err occurs');
});
eventEmitter.emit('error', new Error('programme exception'));
```

运行输出结果：

```
eventEmitter err occurs
```

4.10 实战演练 RESTful API

REST（Representational State Transfer，表述性状态传递）是 Roy Fielding 博士在 2000 年的博士论文中提出来的一种软件架构风格。表述性状态转移是一组架构约束条件和原则，满足这些约束条件和原则的应用程序或设计就是 RESTful。

> **注 意**
>
> REST 是设计风格而不是标准。REST 通常基于使用 HTTP、URI、XML（标准通用标记语言下的一个子集）以及 HTML（标准通用标记语言下的一个应用）这些现有的广泛流行的协议和标准。REST 通常使用 JSON 数据格式。

编写 REST API，实际上就是编写处理 HTTP 请求的 async 函数。HTTP 协议定义了以下 8 种标准的方法：

- GET：请求获取指定资源。
- HEAD：请求指定资源的响应头。
- POST：向指定资源提交数据。
- PUT：请求服务器存储一个资源。
- DELETE：请求服务器删除指定资源。
- TRACE：回显服务器收到的请求，主要用于测试或诊断。
- CONNECT：HTTP/1.1 协议中预留给能够将连接改为管道方式的代理服务器。
- OPTIONS：返回服务器支持的 HTTP 请求方法。

以下为 REST 基本架构的 4 个方法：

- GET：用于获取数据。
- PUT：用于更新或添加数据。
- DELETE：用于删除数据。
- POST：用于添加数据。

REST 请求和普通的 HTTP 请求有几个特殊的地方：

- REST 请求仍然是标准的 HTTP 请求，但是，除了 GET 请求外，POST、PUT 等请求的 body 是 JSON 数据格式，请求的 Content-Type 为 application/json。
- REST 响应返回的结果是 JSON 数据格式，因此，响应的 Content-Type 也是 application/json。

REST 规范定义了资源的通用访问格式，虽然它不是一个强制要求，但遵守该规范可以让人易于理解。

例如，用户信息的处理中，获取所有 User 的 URL 如下：

```
GET /api/listUsers
```

获取某个指定的 User，例如，id 为 100 的 User，其 URL 如下：

```
GET /api/users/100
```

使用 POST 请求新建一个 User，JSON 数据包含在 body 中，URL 如下：

```
POST /api/users
```

使用 PUT 请求更新一个 User，例如，更新 id 为 100 的 User，其 URL 如下：

```
PUT /api/users/100
```

使用 DELETE 请求删除一个 User，例如，删除 id 为 100 的 User，其 URL 如下：

```
DELETE /api/users/100
```

【示例 4-33】

（1）首先，创建一个 JSON 数据资源文件 users.json，内容如下：

```
{
  "user1" : {
    "name" : "mahesh",
    "password" : "password1",
    "profession" : "teacher",
    "id": 1
  },
  "user2" : {
    "name" : "suresh",
    "password" : "password2",
    "profession" : "librarian",
    "id": 2
  },
  "user3" : {
    "name" : "ramesh",
    "password" : "password3",
```

```
      "profession" : "clerk",
      "id": 3
   }
}
```

(2)读取用户信息列表的代码示例如下:

```
var express = require('express');
var app = express();
var fs = require("fs");

app.get('/listUsers', function (req, res) {
   fs.readFile( __dirname + "/" + "users.json", 'utf8', function (err, data) {
       console.log( data );
       res.end( data );
   });
})

var server = app.listen(8081, function () {

  var host = server.address().address
  var port = server.address().port

  console.log("应用实例,访问地址为 http://%s:%s", host, port)

})
```

(3)接下来执行以下命令:

```
$ node server.js
```

(4)在浏览器中访问 http://127.0.0.1:8081/listUsers,结果如下:

```
{
   "user1" : {
      "name" : "mahesh",
      "password" : "password1",
      "profession" : "teacher",
      "id": 1
   },
   "user2" : {
      "name" : "suresh",
      "password" : "password2",
      "profession" : "librarian",
      "id": 2
   },
   "user3" : {
```

```
      "name" : "ramesh",
      "password" : "password3",
      "profession" : "clerk",
      "id": 3
   }
}
```

(5)创建添加用户数据的 API，代码如下：

```
var express = require('express');
var app = express();
var fs = require("fs");

//添加的新用户数据
var user = {
   "user4" : {
      "name" : "mohit",
      "password" : "password4",
      "profession" : "teacher",
      "id": 4
   }
}

app.get('/addUser', function (req, res) {
   // 读取已存在的数据
   fs.readFile( __dirname + "/" + "users.json", 'utf8', function (err, data) {
       data = JSON.parse( data );
       data["user4"] = user["user4"];
       console.log( data );
       res.end( JSON.stringify(data));
   });
})

var server = app.listen(8081, function () {

  var host = server.address().address
  var port = server.address().port
  console.log("应用实例，访问地址为 http://%s:%s", host, port)

})
```

(6)在浏览器中访问 http://127.0.0.1:8081/addUser，结果如下：

```
{ user1:
   { name: 'mahesh',
```

```
    password: 'password1',
    profession: 'teacher',
    id: 1 },
 user2:
  { name: 'suresh',
    password: 'password2',
    profession: 'librarian',
    id: 2 },
 user3:
  { name: 'ramesh',
    password: 'password3',
    profession: 'clerk',
    id: 3 },
 user4:
  { name: 'mohit',
    password: 'password4',
    profession: 'teacher',
    id: 4 }
}
```

（7）创建用于读取指定用户详细信息的 API，示例代码如下：

```
var express = require('express');
var app = express();
var fs = require("fs");

app.get('/getDetail/:id', function (req, res) {
   // 首先我们读取已存在的用户
   fs.readFile( __dirname + "/" + "users.json", 'utf8', function (err, data) {
       data = JSON.parse( data );
       var user = data["user" + req.params.id]
       console.log( user );
       res.end( JSON.stringify(user));
   });
})

var server = app.listen(8081, function () {

  var host = server.address().address
  var port = server.address().port
  console.log("应用实例，访问地址为 http://%s:%s", host, port)

})
```

（8）在浏览器中访问 http://127.0.0.1:8081/getDetail/2，结果如下：

```
{
   "name":"suresh",
   "password":"password2",
   "profession":"librarian",
   "id":2
}
```

（9）创建 RESTful API deleteUser，用于删除指定用户详细信息的 API，代码如下：

```
var express = require('express');
var app = express();
var fs = require("fs");

var id = 2;

app.get('/deleteUser', function (req, res) {

   // First read existing users.
   fs.readFile( __dirname + "/" + "users.json", 'utf8', function (err, data) {
       data = JSON.parse( data );
       delete data["user" + id];

       console.log( data );
       res.end( JSON.stringify(data));
   });
})

var server = app.listen(8081, function () {

  var host = server.address().address
  var port = server.address().port
  console.log("应用实例，访问地址为 http://%s:%s", host, port)

})
```

（10）在浏览器中访问 http://127.0.0.1:8081/deleteUser，结果如下：

```
{ user1:
   { name: 'mahesh',
     password: 'password1',
     profession: 'teacher',
     id: 1 },
  user3:
   { name: 'ramesh',
     password: 'password3',
     profession: 'clerk',
```

```
    id: 3 }
}
```

上述对应的 RESTful API 列表参见表 4.5。

表 4.5 RESTful API 示例

URI	HTTP 方法	发送内容	结　果
listUsers	GET	空	显示所有用户列表
addUser	POST	JSON 字符串	添加新用户
deleteUser	DELETE	JSON 字符串	删除用户
getDetal:id	GET	空	显示用户详细信息

第 5 章 Node.js调试

随着程序越来越复杂,调试工具的重要性越来越突出。客户端脚本用浏览器调试,那么 Node 脚本该如何调试呢?

其实,在 Node.js 程序开发中,经常需要打印调试日志,用的比较多的是 debug 模块,例如 Express 框架中就有使用。根据项目环境需要打印日志、管理日志,在具体的项目实践中有更多的学问,使用好、管理好 Debug 工具是一门艺术。

5.1 基础调试

对于没有任何调试基础的读者,本节先介绍几个简单的调试 API 的方法。有基础的读者可以直接阅读 5.2 节进阶。

5.1.1 基础 API

console 模块提供了基础的调试功能,日志默认打印到控制台,常用的 API 主要有:

- console.log(msg):普通日志打印。
- console.error(msg):错误日志打印。
- console.info(msg):等同于 console.log(msg)。
- console.warn(msg):等同于 console.error(msg)。

【示例 5-1】示例代码如下:

```
console.log('log: hello');
console.log('log: hello', 'world');
console.log('log: hello %s', 'world');

console.error('error: hello');
console.error('error: hello', 'world');
console.error('error: hello %s', 'world');

// 输出如下
```

```
// log: hello
// log: hello world
// log: hello world
// error: hello
// error: hello world
// error: hello world
```

【示例 5-2】通过 console.time(label)和 console.timeEnd(label)打印出两个时间点之间的时间差,单位是毫秒,示例代码如下:

```
var timeLabel = 'hello'

console.time(timeLabel);

setTimeout(console.timeEnd, 1000, timeLabel);
// 输入出入:
// hello: 1005.505ms
```

【示例 5-3】通过 console.assert(value, message)进行断言。如果 value 不为 true,那么抛出 AssertionError 异常,并中断程序执行。代码如下,第二个断言报错,程序将停止执行:

```
console.assert(true, '1、right');
console.assert(false, '2、right', '2、wrong');

// 输出如下
// assert.js:90
//     throw new assert.AssertionError({
//     ^
//     AssertionError: 2、right 2、wrong
//         at Console.assert (console.js:95:23)
```

为避免程序异常退出,需要对以上异常进行处理,比如:

```
try{
   console.assert(false, 'error occurred');
}catch(e){
   console.log(e.message);
}

// 输出如下
// error occurred
```

【示例 5-4】通过 console.trace(msg)打印错误堆栈,将 msg 打印到标准错误输出流里,包含当前代码的位置和堆栈信息,示例代码如下:

```
console.trace('trace is called');

// 输出如下
```

```
    // Trace: trace is called
    //                                              at        Object.<anonymous>
(/Users/a/Documents/git-code/nodejs-learning-guide/examples/2016.12.01-console
/trace.js:1:71)
    //     at Module._compile (module.js:541:32)
    //     at Object.Module._extensions..js (module.js:550:10)
    //     at Module.load (module.js:456:32)
    //     at tryModuleLoad (module.js:415:12)
    //     at Function.Module._load (module.js:407:3)
    //     at Function.Module.runMain (module.js:575:10)
    //     at startup (node.js:160:18)
    //     at node.js:445:3
```

通过 console.dir(obj)深层打印，例如：

```
var obj = {
    nick: 'chyingp'
};

console.log(obj);  // 输出: { nick: 'chyingp' }
console.dir(obj);  // 输出: { nick: 'chyingp' }
```

当 obj 的层级比较深时，深层打印的作用就比较明显。可以通过 depth 自定义打印的层级数，默认是 2，这在调试时较为有用：

```
var obj2 = {
    human: {
        man: {
            info: {
                nick: 'chyingp'
            }
        }
    }
};

console.log(obj2);  // 输出: { human: { man: { info: [Object] } } }
console.dir(obj2);  // 输出: { human: { man: { info: [Object] } } }

console.dir(obj2, {depth: 3});  // 输出: { human: { man: { info: { nick:
'chyingp' } } } }
```

5.1.2　自定义 stdout

基于 Console 类可以较为方便地扩展出需要的 console 实例，例如试图把调试信息打印到文件里，而不是输出在控制台上，可以通过 new console.Console(stdout, stderr)创建自定义的 console 实例：

```
var fs = require('fs');
var file = fs.createWriteStream('./stdout.txt');

var logger = new console.Console(file, file);

logger.log('hello');
logger.log('word');

// 备注：内容输出到 stdout.txt 里，而不是打印到控制台
```

5.1.3 控制调试日志

通常直接使用 console.log 输出调试日志，使用 console 对象直接将日志输出到控制台，由于 Node.js 和浏览器环境都默认支持 console 对象，因此这种方式是很直接、很简洁的日志调试方法。

但是，随着项目规模不断扩大，console 控制台输出的日志会逐渐堆积，且多得不可读。另一方面，在开发调试环境才需要打开日志，在线上生产环境就不需要输出过多的日志。使用 console 对象控制日志输出需要以"加注释""去注释"的方式来控制日志的输出，这种方式非常低效。

debug 库就是专门控制输出调试日志的库，能够完美解决上述需求。

首先，debug 库会判断 DEBUG 环境变量，因此不需要修改代码，只需要调整程序运行环境就可以控制日志是否输出。另外，debug 库不是简单地布尔判断 DEBUG 环境变量，而是会对 DEBUG 环境变量进行解析，允许开发者选择性地控制输出哪些模块的日志，有效地解决了调试程序时控制台的日志堆积问题。通过控制 debug 可以让其只输出需要关心的程序模块日志信息。

【示例 5-5】以下是使用了 debug 库的程序示例。

首先新建 app_debug.js 文件，输入以下代码：

```
var debug = require('debug')('http')
  , http = require('http')
  , name = 'My App';

// fake app

debug('booting %o', name);

http.createServer(function(req, res){
  debug(req.method + ' ' + req.url);
  res.end('hello\n');
}).listen(3000, function(){
  debug('listening');
```

```
});

// fake worker of some kind

require('./worker');
```

新建 worker.js 文件,输入以下代码:

```
var a = require('debug')('worker:a')
  , b = require('debug')('worker:b');

function work() {
  a('doing lots of uninteresting work');
  setTimeout(work, Math.random() * 1000);
}

work();

function workb() {
  b('doing some work');
  setTimeout(workb, Math.random() * 2000);
}

workb();
```

如果设置环境变量 DEBUG 为 worker:*,就会输出所有的日志,运行 DEBUG=worker:* node app_debug.js 命令,可以看到如图 5.1 所示的输出。

图 5.1 使用 debug 模块输出所有日志

如果设置环境变量 DEBUG 为 worker:a,就只会输出 worker:a 的日志。运行 DEBUG=worker:a node app_debug.js,可以看到如图 5.2 所示的日志输出。

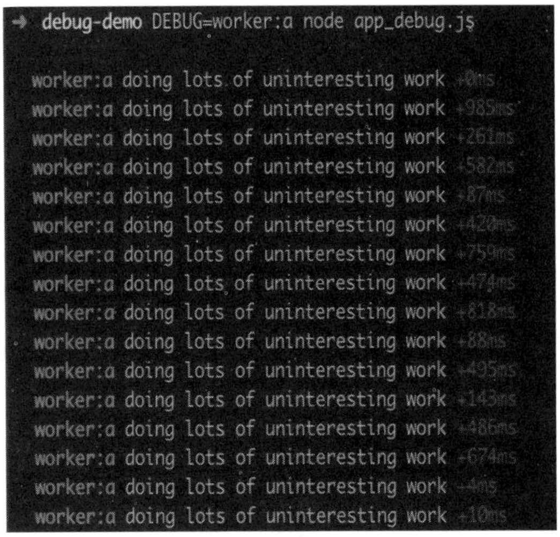

图 5.2　使用 debug 输出部分日志

从上述例子中得知 DEBUG 环境变量的设置支持通配符*。假设程序中存在调试器：connect:bodyParser、connect:compress、connect:session，可以将 DEBUG 设置为 DEBUG=connect:bodyParser,connect:compress,connect:session，或者简单地使用通配符 DEBUG=connect:*。如果需要调试非 connect 相关的其他信息，就可以使用 "-" 符号，"减去" connect：DEBUG=*,-connect:* 。

在浏览器端，debug 也能够很好地运行，但是在浏览器端开启调试并不是使用 DEBUG 环境变量，因为浏览器端不支持 process.env.DEBUG 访问。在浏览器端，使用 localStorage 对象控制 debug：

```
localStorage.debug = 'worker:*'
```

在实际项目中，往往线下环境要开启调试，而线上环境是否需要开启调试是有条件的，示例代码如下：

```
const _ = require('lodash');
const debug = require('debug');
const debugA = debug('A:');
const debugB = debug('B:');

// 当环境为 production 时，所有的 debugA 均不会输出

if (process.env.NODE_ENV === 'production') {
  debugA.enabled = false;
}

debugA('hello world');
debugB('I am new to debug');
```

一个项目通常会有开发版本、测试版本以及线上版本，每个版本可能都会对应不同的相关

参数，或许数据库的连接地址不同，或许请求的 API 地址不同，等等。为了方便管理，通常做成配置文件的形式，根据不同的环境加载不同的文件。通过环境变量NODE_ENV切换环境，日志输出效果如图 5.3 所示。

```
→ debug-demo DEBUG=A: node app_production.js
A: hello world
→ debug-demo DEBUG=A: NODE_ENV=production node app_production.js
→ debug-demo DEBUG=B: NODE_ENV=production node app_production.js
B: I am new to debug +0ms
```

图 5.3　根据环境判断是否输出日志

当项目程序变得复杂时，我们需要对日志进行分类打印，debug 支持命名空间，说明如下：

- DEBUG=app,api：表示同时打印出命名空间为 app、api 的调试日志。
- DEBUG=a*：支持通配符，所有命名空间为 a 开头的调试日志都打印。

```
/**
 * debug 例子：命名空间
 */
var debug = require('debug');
var appDebug = debug('app');
var apiDebug = debug('api');

// 分别运行下面几行命令查看效果
//
//     DEBUG=app node 02.js
//     DEBUG=api node 02.js
//     DEBUG=app,api node 02.js
//     DEBUG=a* node 02.js
//
appDebug('hello');
apiDebug('hello');
```

debug 支持自定义格式化，包括命名空间和色彩定义，函数性能比较的代码示例如下：

```
// 计算函数性能示例，传入一个纯数字数组，计算其平均值
// avgA 使用 lodash 库中的 sum 求和
// avgB 使用 ES5中的 Array.proptype.reduce 求和
const _ = require('lodash');
const debug = require('debug');
const debugA = debug('avgA:');
const debugB = debug('avgB:');
const testFixture = [0, 1, 2, 3, 4, 5, 6, 7, 8, 9, 10];

// 使用 lodash sum 求和
```

```
function avgA(arr) {
  return _.sum(arr) / arr.length;
}

// 使用原生 reduce 求和
function avgB(arr) {
  return arr.reduce((cal, curr) => cal + curr, 0) / arr.length;
}

const range = _.range(0, 10000000, 1);
range.forEach((index) => {
  const result = avgA(testFixture);
  if (index % 1000000 === 0) {
    debugA('time %d', index);
  }
})

range.forEach((index) => {
  const result = avgB(testFixture);
  if (index % 1000000 === 0) {
    debugB('time %d', index);
  }
})
```

运行 DEBUG=* NODE_ENV=production node app_avg.js 命令，效果如图 5.4 所示。

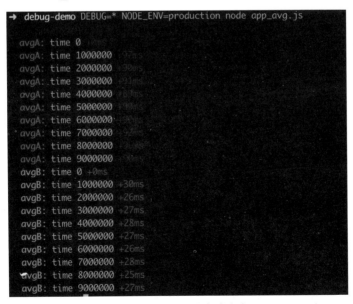

图 5.4 debug 支持色彩定义

5.2 进阶调试

目前,Node.js 断点调试的主要方式有通过 inspector、通过 IDE（如 vscode 编辑器等,其本质原理是类似的。

【示例 5-6】首先,准备需要调试的示例代码。

新建一个工作目录,并进入该目录:

```
$ mkdir debug-demo
$ cd debug-demo
```

然后,生成 package.json 文件,并安装 Koa 框架和 koa-route 模块:

```
$ npm init -y
$ npm install --save koa koa-route
```

接着,新建一个脚本 app.js,并写入下面的内容:

```javascript
// app.js
const Koa = require('koa');
const router = require('koa-route');

const app = new Koa();

const greet = 'Hello World';
const country = 'China';

const str = greet + ' live in ' + country;

const main = ctx => {
  ctx.response.body = 'Hello World';
};

const welcome = (ctx, name) => {
  ctx.response.body = 'Hello ' + name;
};

app.use(router.get('/', main));
app.use(router.get('/:name', welcome));

app.listen(3000);
console.log('listening on port 3000');
```

5.2.1 使用 Inspect 调试

前面简单介绍过 Inspect，这里再重温一下。

使用如下命令运行上面的脚本启动开发者工具：

```
$node--inspect app.js
```

上面的命令中，--inspect 参数是启动调试模式必需的。--inspect 和--inspect-brk 的区别是，--inspect-brk 默认会在第一行代码设置断点。这时，打开浏览器访问 http://127.0.0.1:3000，就可以看到"Hello World"，如图 5.5 所示。

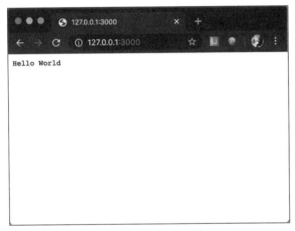

图 5.5　使用--inspect 运行程序

接下来，就要开始调试了。有两种打开调试工具的方法，我们依次介绍。

（1）第一种方法，在 Chrome 浏览器的地址栏输入 chrome://inspect 或者 about:inspect，按回车键后可以看到如图 5.6 所示的界面。

图 5.6　Chrome inspect 界面

在 Target 部分，单击 inspect 链接，进入调试工具，如图 5.7 所示。

图 5.7　inspect 调试界面

（2）第二种方法，在 http://127.0.0.1:3000 的窗口打开开发者工具，顶部左上角有一个 Node 的绿色标志，单击该图标即可进入调试界面，如图 5.8 所示。

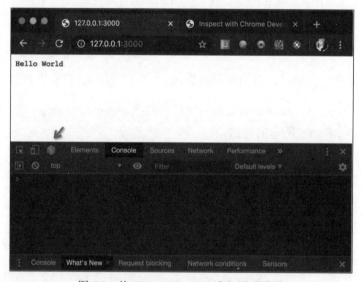

图 5.8　从 Chrome dev tool 进入调试界面

我们调试的时候并不希望调试到基础库（jQquery、Node）的代码，调试时可以屏蔽基础库代码。Chrome 提供了 Blackbox 功能。Blackbox 允许屏蔽指定 JS 文件，这样调试就可以绕过这些文件，如图 5.9~图 5.11 所示。

图 5.9　Blackbox 示例

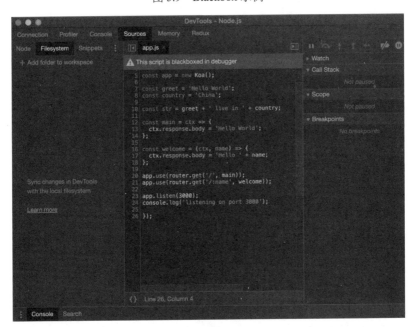

图 5.10　开启 Blackbox

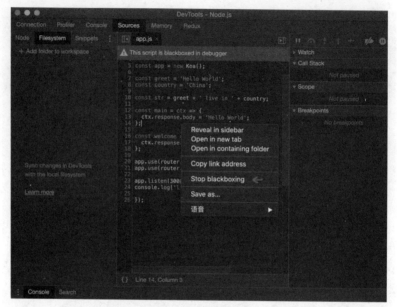

图 5.11　关闭 Blackbox

5.2.2　使用 VSCode IDE 调试

本节介绍使用 VSCode 调试 Node.js。

（1）首先，在 VSCode 里打开项目，单击界面左侧第 4 个调试并运行按钮，如图 5.12 所示。

图 5.12　在 VSCode 中调试

（2）单击调试按钮，选择 Node.js，如图 5.13 所示。

图 5.13 选择 Node.js

（3）单击设置按钮，如图 5.14 所示。

图 5.14 设置调试配置

（4）选择完成之后会生成一个.vscode 文件夹，文件夹下有一个 launch.json 文件。添加调试配置主要需要修改的是可执行文件的路径。将 program 字段的值修改为自己程序的入口文件，开始调试时会从该入口启动程序，示例项目的入口为 app.js，修改为如图 5.15 所示的样子。

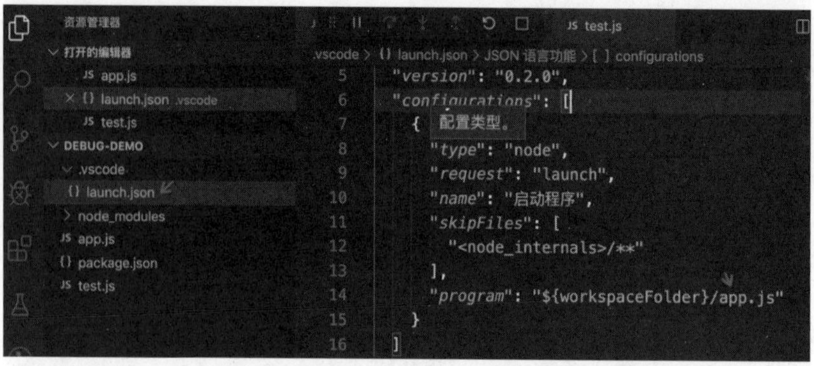

图 5.15 配置调试入口

（5）修改完成后，单击绿色三角按钮，设置断点，开始调试程序，如图 5.16 所示。

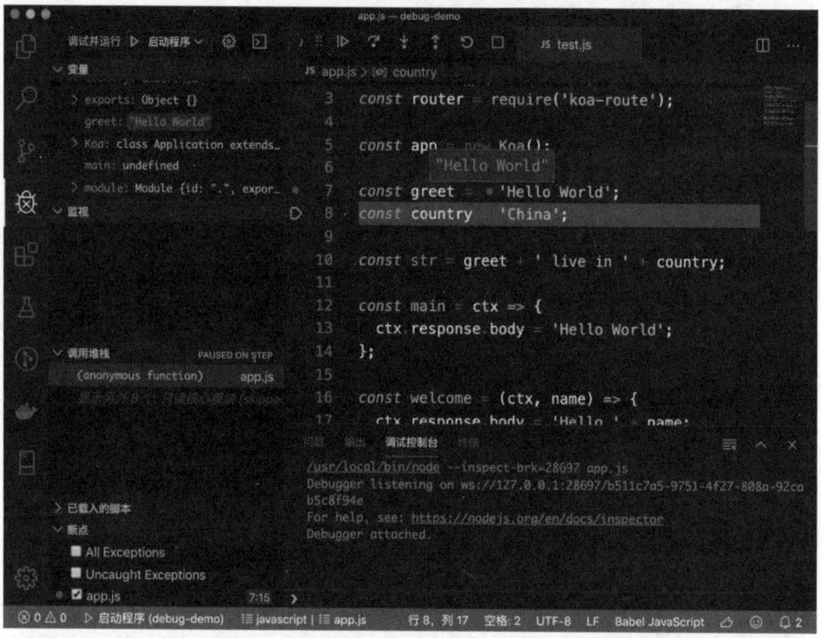

图 5.16 设置断点调试

从图 5.16 可以看出顺利断点，左侧可以看到有变量、监视对象，右侧的调试工具栏与 Chrome Devtools 类似。

第 6 章 Node.js 的异步 I/O 与多线程

Node.js 具有以下特点：

- 异步 I/O（非阻塞 I/O）
- 事件驱动
- 单线程

单线程指的是主线程是"单线程"的，所有阻塞的部分交给一个线程池处理，然后这个主线程通过一个队列跟线程池协作，于是对我们写到的 JavaScript 代码部分，不再关心线程问题，代码也主要由一堆 callback 回调构成，然后主线程在不停的循环过程中适时调用这些代码。Node.js 采用单线程异步非阻塞模式，也就是说每一个计算独占 CPU，遇到 I/O 请求不阻塞后面的计算，当 I/O 完成后，以事件的方式通知，继续执行计算。Node.js 擅长 I/O 密集型应用，不擅长 CPU 密集型应用。

6.1 异步 I/O

在操作系统中，程序运行的空间分为内核空间和用户空间。我们常常提起的异步 I/O，其实质是用户空间中的程序不用依赖内核空间中的 I/O 操作实际完成，即可进行后续任务。以下伪代码模仿了一个从磁盘上获取文件和一个从网络中获取文件的操作。异步 I/O 的效果就是 getFileFromNet 的调用不依赖于 getFile 调用的结束。

```
getFile("file_path");
getFileFromNet("url");
```

如果以上两个任务的时间分别为 m 和 n，采用同步方式的程序要完成这两个任务的时间总花销就会是 m + n。但是如果是采用异步方式的程序，在两种 I/O 可以并行的状况下，比如网络 I/O 与文件 I/O，时间开销就会减小为 max(m, n)。

6.1.1 异步 I/O 的必要性

有的编程语言为了设计得使应用程序调用方便，将程序设计为同步 I/O 的模型。这意味着程序中的后续任务都需要等待 I/O 的完成。在等待 I/O 完成的过程中，程序无法充分利用 CPU。为了充分利用 CPU 和使 I/O 可以并行，目前有两种方式可以实现多线程单进程和单线程多进程。

1. 多线程单进程

多线程的设计就是为了在共享的程序空间中实现并行处理任务，从而达到充分利用 CPU 的效果。多线程的缺点在于执行时上下文交换的开销较大以及状态同步（锁）的问题。同样它也使得程序的编写和调用复杂化。

2. 单线程多进程

为了避免多线程造成的使用不便问题，有的语言选择了单线程保持调用简单化，采用启动多进程的方式来达到充分利用 CPU 和提升总体的并行处理能力。它的缺点在于业务逻辑复杂时（涉及多个 I/O 调用），因为业务逻辑不能分布到多个进程之间，事务处理时长要远远大于多线程模式。

前者在性能优化上还有回旋的余地，后者的做法纯粹是一种增加服务器的行为。

而且现在的大型 Web 应用中，单机的情形是十分稀少的，一个事务往往需要跨越网络几次才能完成最终处理。如果网络速度不够理想，m 和 n 值都将会变大，这时同步 I/O 的语言模型就会暴露出其脆弱的状态。

这种场景下的异步 I/O 将会体现其优势，max(m, n)的时间开销可以有效地缓解 m 和 n 值增长带来的性能问题。而当并行任务更多的时候，m + n + ⋯与 max(m, n, ⋯) 之间孰优孰劣更是一目了然。从这个公式中，可以了解到异步 I/O 在分布式环境中是多么重要，而 Node.js 天然地支持这种异步 I/O，这是众多云计算厂商对其青睐的根本原因。

6.1.2 操作系统对异步 I/O 的支持

我们说到 Node.js 时，常常会将异步、非阻塞、回调、事件这些词语混合在一起。其中，异步与非阻塞听起来似乎是同一回事。从实际效果的角度来说，异步和非阻塞都达到了并行 I/O 的目的。但是对计算机内核 I/O 而言，异步/同步和阻塞/非阻塞实际上是两回事。同步与异步是对应的，它们是线程之间的关系，两个线程之间要么是同步的，要么是异步的。阻塞与非阻塞是对同一个线程来说的，在某个时刻，线程要么处于阻塞状态，要么处于非阻塞状态。阻塞是使用同步机制的结果，非阻塞则是使用异步机制的结果。

1. I/O 的阻塞与非阻塞

阻塞模式的 I/O 会造成应用程序等待，直到 I/O 完成。同时，操作系统也支持将 I/O 操作设置为非阻塞模式，这时应用程序的调用将可能在没有拿到真正数据时就立即返回了，为此应用程序需要多次调用才能确认 I/O 操作完全完成。

2. I/O 的同步与异步

I/O 的同步与异步出现在应用程序中。如果做阻塞 I/O 调用，应用程序等待调用完成的过程就是一种同步状况。相反，I/O 为非阻塞模式时，应用程序则是异步的。

6.1.3 异步 I/O 与轮询技术

当进行非阻塞 I/O 调用时，要读到完整的数据，应用程序需要进行多次轮询，才能确保读取数据完成，以进行下一步的操作。

轮询技术的缺点在于应用程序要主动调用，会造成占用较多 CPU 时间片,性能较为低下。现存的轮询技术有以下这些：

- read
- select
- poll
- epoll
- pselect
- kqueue

read 是性能最低的一种，它通过重复调用来检查 I/O 的状态以完成完整数据的读取。select 是一种改进方案，通过对文件描述符上的事件状态来进行判断。操作系统还提供了 poll、epoll 等多路复用技术来提高性能。

轮询技术满足了异步 I/O 确保获取完整数据的保证。但是对于应用程序而言，它仍然只能算是一种同步，因为应用程序仍然需要主动去判断 I/O 的状态，依旧花费了很多 CPU 时间来等待。

重复调用 read 进行轮询直到最终成功，用户程序会占用较多 CPU，性能较为低下。而实际上操作系统提供了 select 方法来代替这种重复 read 轮询进行状态判断。select 内部通过检查文件描述符上的事件状态来判断数据是否完全读取。但是对于应用程序而言，它仍然只能算是一种同步，因为应用程序仍然需要主动去判断 I/O 的状态，依旧花费了很多 CPU 时间等待，select 也是一种轮询。

6.2 进程、线程、协程等

本节介绍进程、线程、协程，以及并发、并行的概念。

6.2.1 进程、线程、协程

首先介绍进程、线程、协程的概念。

1. 进程

进程的出现是为了更好地利用 CPU 资源使得并发成为可能。假设有两个任务 A 和 B，当 A 遇到 IO 操作时，CPU 默默地等待任务 A 读取完操作再去执行任务 B，这样无疑是对 CPU 资源的极大浪费。有人就在想，可以在任务 A 读取数据时让任务 B 执行，当任务 A 读取完数据后，再切换到任务 A 执行。

注意关键字"切换"，这就涉及状态的保存和状态的恢复，加上任务 A 与任务 B 所需要的系统资源（内存、硬盘、键盘等）是不一样的。自然就需要有一个东西去记录任务 A 和任务 B 分别需要什么资源，怎样去识别任务 A 和任务 B，等等。进程就被发明出来了。通过进程来分配系统资源，标识任务。如何分配 CPU 去执行进程称为调度，进程状态的记录、恢复、切换称为上下文切换。进程是系统资源分配的最小单位，进程占用的资源有：地址空间、全局变量、文件描述符、各种硬件等资源。

2. 线程

线程的出现是为了降低上下文切换的消耗，提高系统的并发性，并突破一个进程只能干一样事的缺陷，使得进程内并发成为可能。假设一个文本程序需要接收键盘输入，将内容显示在屏幕上，还需要保存信息到硬盘中。若只有一个进程，势必造成同一时间只能干一样事的尴尬（当保存时，就不能通过键盘输入内容）。若有多个进程，则每个进程负责一个任务，进程 A 负责接收键盘输入的任务，进程 B 负责将内容显示在屏幕上的任务，进程 C 负责保存内容到硬盘中的任务。这里进程 A、B、C 间的协作涉及进程通信问题，而且有共同需要拥有的东西——文本内容，不停地切换造成性能上的损失。如果有一种机制可以使任务 A、B、C 共享资源，上下文切换所需要保存和恢复的内容就少了，同时又可以减少通信所带来的性能损耗。

这种机制就是线程。线程共享进程的大部分资源，并参与 CPU 的调度，当然线程也是拥有自己的资源的，例如栈、寄存器等。此时，进程同时也是线程的容器。线程也有自己的缺陷，例如健壮性差，如果一个线程挂掉了，整个进程也就挂掉了，这意味着其他线程也挂掉了，进程却没有这个问题，一个进程挂掉，其他的进程还是运行着的。

3. 协程

协程通过在线程中实现调度避免了陷入内核级别的上下文切换造成的性能损失，进而突破了线程在 IO 上的性能瓶颈。当涉及大规模的并发连接时，例如 10KB 连接，以线程作为处理单元，系统调度的开销还是过大。当连接数很多时，需要大量的线程来干活，可能大部分线程处于 ready 状态，系统会不断地进行上下文切换。既然性能瓶颈在上下文切换，解决思路也就有了，在线程中自己实现调度，不陷入内核级别的上下文切换。说明一下，在历史上，协程比线程要出现得早，在 1963 年首次提出，但没有流行开来。

6.2.2 应用场景

进程、线程、协程不断突破，以更高效地处理阻塞，不断地提高 CPU 的利用率。但是并不是说，线程就一定比进程快，而协程就一定比线程快，具体还是要看应用场景。可以简单地把应用分为 IO 密集型应用和 CPU 密集型应用。

1. 多核 CPU，CPU 密集型应用

此时多线程的效率是最高的，多线程可以使得全部 CPU 核心满载，又避免了协程间切换造成性能损失。在 CPU 密集型任务中，CPU 一直在利用着，切换反而会造成性能损失，即便协程上下文切换消耗最小，但也是有消耗的。

2. 多核 CPU，IO 密集型应用

此时采用多线程多协程效率最高，多线程可以使得全部 CPU 核心满载，而一个线程多协程则更好地提高了 CPU 的利用率。

3. 单核 CPU，CPU 密集型应用

此时单进程效率最高，单个进程已经使得 CPU 满载了。

4. 单核 CPU，IO 密集型应用

此时多协程效率最高。

6.2.3 并发与并行

首先介绍并发与并行的概念。

1. 并行

并行就是指同一时刻有两个或两个以上的"工作单位"在同时执行，从硬件的角度上来看，就是同一时刻有两条或两条以上的指令处于执行阶段。所以，多核是并行的前提，单线程永远无法达到并行状态。可以利用多线程和多进程达到并行状态。另外，由于 GIL 的存在，对于 Python 来说无法通过多线程达到并行状态。

2. 并发

对于并发，要从两方面去理解：

（1）并发设计

（2）并发执行

先介绍并发设计，当说一个程序是并发的时候，更多的是指这个程序采取了并发设计。并发设计的标准是：使多个操作可以在重叠的时间内进行。这里的重点在于重叠的时间内，重叠

的时间可以理解为一段时间内。例如，在 1 秒内，具有 IO 操作的 task1 和 task2 都完成，就可以说是并发执行的。所以，单线程也是可以做到并发执行的。当然，并行肯定是并发的。一个程序能否并发执行取决于设计，也取决于部署方式。例如，当给程序开一个线程（协程是不开的）时，它不可能是并发的，因为在重叠的时间内根本就没有两个 task 在运行。当一个程序被设计成完成一个任务再去完成下一个任务的时候，即便部署是多线程多协程的，也是无法达到并发执行的。

并行与并发的关系是：并发的设计使得并发执行成为可能，而并行是并发执行的其中一种模式。

6.3 在 Node.js 中实现多线程

Google 的 V8 JavaScript 引擎已经在 Chrome 浏览器里证明了它的性能，所以 Node.js 的作者 Ryan Dahl 选择了 V8 作为 Node.js 的执行引擎，V8 赋予 Node.js 高效性能的同时也注定了 Node.js 和大名鼎鼎的 Nginx 一样，都是以单线程为基础的，当然这也正是作者 Ryan Dahl 设计 Node.js 的初衷。

那么多线程又是怎么回事呢？本节来揭晓答案。

6.3.1 单线程的 JavaScript

Node.js 的单线程具有它的优势，但也并非十全十美，在保持单线程模型的同时，它是如何保证非阻塞的呢？

1. 高性能

首先，单线程避免了传统 PHP 那样频繁创建、切换进程的开销，使得执行速度更加迅速。

其次，资源占用少，对 Node.js 的 Web 服务器做过压力测试的朋友可能会发现，Node.js 在大负荷下对内存的占用仍然很低，同样的负载，PHP 因为一个请求一个线程的模型，将会占用大量的物理内存，很可能会导致服务器因物理内存耗尽而频繁交换，失去响应。

2. 线程安全

单线程的 JavaScript 还保证了绝对的线程安全，不用担心同一变量同时被多个线程进行读写而造成的程序崩溃。例如 Web 访问统计时，因为单线程的绝对线程安全，所以不可能存在同时对 count 变量进行读写的情况，我们的统计代码即使是成百的并发用户请求都不会出现问题，相较于 PHP 的那种存文件记录访问，就会面临并发同时写文件的问题。

线程安全的同时也解放了开发人员，免去了多线程编程中忘记对变量加锁或者解锁造成的悲剧。

3. 单线程的异步和非阻塞

Node.js 是单线程的，它是如何做到 I/O 的异步和非阻塞的呢？其实 Node.js 在底层访问 I/O 还是多线程的，有兴趣的朋友可以翻看 Node.js 的 fs 模块的源码，里面用到 libuv 来处理 I/O，所以在我们看来 Node.js 的代码就是非阻塞和异步形式的。

4. 阻塞的单线程

既然 Node.js 是单线程异步非阻塞的，我们是否就可以高枕无忧了呢？在浏览器中，JavaScript 都是以单线程的方式运行的，所以我们不用担心 JavaScript 同时执行带来的冲突问题，这对于我们的编码带来很多便利。

但是对于在服务端执行的 Node.js，它可能每秒有上百个请求需要处理，对于在浏览器端工作良好的单线程，JavaScript 是否能同样在服务端表现良好呢？

【示例 6-1】我们看如下代码：

```
var start = Date.now();//获取当前时间戳
setTimeout(function () {
    console.log(Date.now() - start);
    for (var i = 0; i < 1000000000; i++){//执行长循环
    }
}, 1000);
setTimeout(function () {
    console.log(Date.now() - start);
}, 2000);
```

最终的打印结果是（结果可能因为你的机器配置的不同而不同）：

```
1000
3738
```

对于我们期望 2000 毫秒后执行的 setTimeout 函数，其实经过了 3738 毫秒之后才执行，换而言之，因为执行了一个很长的 for 循环，所以我们整个 Node.js 主线程都被阻塞了，如果在处理 100 个用户请求时，其中第一个就需要这样大量的计算，那么其余 99 个都会被延迟执行。

其实虽然 Node.js 可以处理数以千记的并发，但是一个 Node.js 进程在某一时刻其实只是在处理一个请求。

5. 单线程和多核

线程是 CPU 调度的一个基本单位，一个 CPU 同时只能执行一个线程的任务，同样一个线程任务也只能在一个 CPU 上执行，所以如果运行 Node.js 的机器配置的是像 i5、i7 这样的多核 CPU，那么将无法充分利用多核 CPU 的性能来为 Node.js 服务。

6.3.2　Node.js 内部分层

从图 6.1 可以看到，Node.js 的结构大致分为 3 个层次：

（1）Node.js 标准库，这部分是由 JavaScript 编写的，即我们在使用过程中能够直接调用的 API。在源码中的 lib 目录下可以看到。

（2）Node Bindings，这一层是 JavaScript 与底层 C/C++能够沟通的关键，前者通过 bindings 调用后者，相互交换数据。

（3）这一层是支撑 Node.js 运行的关键，由 C/C++ 实现，其中每一项说明如下：

- V8：Google 推出的 JavaScript VM，也是 Node.js 使用 JavaScript 的关键原因，它为 JavaScript 提供了在非浏览器端运行的环境，它的高效是 Node.js 之所以高效的原因之一。
- libuv：它为 Node.js 提供了跨平台、线程池、事件池、异步 I/O 等能力，是 Node.js 如此强大的关键。
- C-ares：提供了异步处理 DNS 相关的能力。
- http_parser、OpenSSL、zlib 等：提供包括 HTTP 解析、SSL、数据压缩等其他的能力。

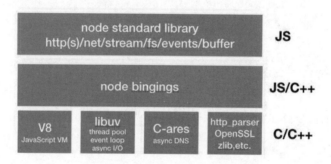

图 6.1　Node.js 内部分层结构

6.3.3　libuv

如果我们非要让 Node.js 支持多线程，还是提倡使用官方的做法，利用 libuv 库来实现。

libuv 是一个跨平台的异步 I/O 库，它主要用于 Node.js 的开发，同时也被 Mozilla's Rust Language、Luvit、Julia、pYUV 等使用。它主要包括 Event Loops（事件循环）、Filesystem（文件系统）、Networking（网络支持）、Threads（线程）、Processes（进程）、Utilities（其他工具）。

在 Node.js 核心 API 中，异步多线程大多是使用 libuv 来实现的。从图 6.2 可以看出，几乎所有和操作系统打交道的部分都离不开 libuv 的支持。libuv 也是 Node 实现跨操作系统的核心所在。

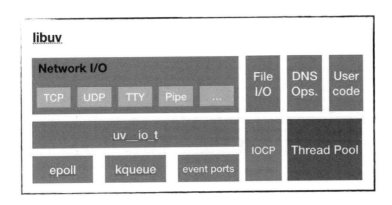

图 6.2　libuv

6.3.4　多进程

在支持 HTML5 的浏览器里，我们可以使用 Web Worker 来将一些耗时的计算丢入 worker 进程中执行，这样主进程就不会阻塞，用户也就不会有卡顿的感觉了。在 Node.js 中是否也可以使用这类技术来保证主线程的畅通呢？

1. cluster

cluster 可以用来让 Node.js 充分利用多核 CPU 的性能，同时也可以让 Node.js 程序更加健壮，官网上的 cluster 示例已经告诉我们如何重新启动一个因为异常而奔溃的子进程。

2. Web Worker

想要像在浏览器端那样启动 worker 进程，我们需要利用 Node.js 核心 API 里的 child_process 模块。child_process 模块提供了 fork 的方法，可以启动一个 Node.js 文件，将它作为 worker 进程，当 worker 进程工作完毕后，把结果通过 send 方法传递给主进程，然后自动退出，这样我们就可以利用多进程来解决主线程阻塞的问题。

【示例 6-2】我们先启动一个 Web 服务，通过接收参数 n 来计算斐波那契数组：

```
var express = require('express');
var fork = require('child_process').fork;
var app = express();
app.get('/', function(req, res){
  var worker = fork('./work_fibo.js')    //创建一个工作进程
  worker.on('message', function(m) {        //接收工作进程计算结果
        if('object' === typeof m && m.type === 'fibo'){
            worker.kill();              //发送杀死进程的信号
            res.send(m.result.toString());//将结果返回客户端
        }
  });
  worker.send({type:'fibo',num:~~req.query.n || 1});
  //发送给工作进程计算 fibo 的数量
});
app.listen(8124);
```

我们通过 Express 监听 8124 端口，对每个用户的请求都会去派生一个子进程，通过调用 worker.send 方法将参数 n 传递给子进程，同时监听子进程发送消息的 message 事件，将结果响应给客户端。

下面是被 fork 的 work_fibo.js 文件内容：

```
var fibo = function fibo (n) {//定义算法
    return n > 1 ? fibo(n - 1) + fibo(n - 2) : 1;
}
process.on('message', function(m) {
//接收主进程发送过来的消息
        if(typeof m === 'object' && m.type === 'fibo'){
            var num = fibo(~~m.num);
            //计算jibo
            process.send({type: 'fibo',result:num})
            //计算完毕返回结果
        }
});
process.on('SIGHUP', function() {
    process.exit();//收到kill信息，进程退出
});
```

我们先定义函数 fibo 用来计算斐波那契数组，然后监听主线程发来的消息，计算完毕之后将结果发送到主线程。同时还监听 process 的 SIGHUP 事件，触发此事件就退出进程。

> **注 意**
>
> 主线程的 kill 方法并不是真的使子进程退出，而是会触发子进程的 SIGHUP 事件，真正的退出还是依靠 "process.exit();"。

6.4 Node 性能小结

单线程的 Node.js 给我们编码带来了太多的便利和乐趣，但要注意任何一个隐藏的问题都可能击溃整个线上正在运行的 Node.js 程序。

单线程异步的 Node.js 不代表不会阻塞，在主线程做过多的任务时可能会导致主线程卡死，从而影响整个程序的性能，所以我们要非常小心地处理大量的循环、字符串拼接和浮点运算等 CPU 密集型任务，合理地利用各种技术把任务丢给子线程或子进程去完成，保持 Node.js 主线程的畅通。

线程/进程的使用并不是没有开销的，尽可能减少创建和销毁线程/进程的次数可以提升系统整体的性能并降低出错的概率。

最后，不要一味地追求高性能和高并发，因为我们可能不需要系统具有那么大的吞吐率。高效、敏捷、低成本的开发才是项目所需要的，这也是 Node.js 能够在众多开发语言中脱颖而出的关键。

第 7 章
◀ Node.js的错误处理 ▶

异常处理是程序运行中必须关注的。当异常出现后,应该第一时间关注到,并且快速解决。如果我们不处理异常的话,直接就会导致程序崩溃,用户体验比较差,因此在写代码时就要对异常提前做预防处理,尽量保证在异常出现时给用户一个友好的提示,不至于服务挂起导致请求超时,并且能将异常信息记录并上报,方便后期排查解决。

7.1 错误的分类

错误分成两大类:操作失败和程序员的失误。操作失败是正确编写的程序在运行时产生的错误。它并不是程序的 Bug,反而经常是其他问题:系统本身的问题(内存不足或者打开的文件数过多)、系统配置的问题(没有到达远程主机的路由)、网络问题(端口挂起)、远程服务(500 错误,连接失败)等,这些问题主要是:

- 连接不到服务器。
- 无法解析主机名。
- 无效的用户输入。
- 请求超时。
- 服务器返回 500。
- 套接字被挂起。
- 系统内存不足。

程序员失误是系统的 Bug,这些错误往往可以通过修改代码来避免,没法被有效地处理,例如:

- 读取 undefined 的一个属性。
- 调用异步函数没有指定回调。
- 应该传对象的时候传了一个字符串。
- 应该传 IP 地址的时候传了一个对象。

错误处理不能凭空加到一个没有任何错误处理的程序中。用户没有办法在一个集中的地方处理所有的异常,就像用户不能在一个集中的地方解决所有的性能问题。用户需要考虑任何会导致失败的代码(比如打开文件、连接服务器、Fork 子进程等)可能产生的结果,包括为什么出错，错误背后的原因。关键在于错误处理的粒度要细，因为哪里出错和为什么出错决定了影响大小和对策。

错误是 Error 的一个实例。错误被创建并且直接传递给另一个函数或者被抛出。如果一个错误被抛出了，它就变成了一个异常，例如：

```
throw new Error('something bad happened');
```

但是使用一个错误而不抛出也是可以的：

```
callback(new Error('something bad happened'));
```

这种用法更常见，因为在 Node.js 里，大部分的错误都是异步的。实际上，try/catch 唯一常用的是在 JSON.parse 和类似验证用户输入的地方。

7.2 函数的错误处理

函数有三种基本的传递错误的模式：

（1）throw。throw 以同步的方式传递异常。也就是在函数被调用处的相同的上下文。如果调用者（或者调用者的调用者）用了 try/catch，异常就可以被捕获。如果所有的调用者都没有用，那么程序通常会崩溃（异常也可能会被 domains 或者进程级的 uncaughtException 捕捉到，详见下文）。

（2）callback。callback 是基础的异步传递事件的一种方式。用户传进来一个函数（callback），之后当某个异步操作完成后调用这个 callback。通常 callback 会以 callback(err,result)的形式被调用，这种情况下，err 和 result 必然有一个是非空的，取决于操作是成功还是失败。

（3）EventEmitter。更复杂的情形是，函数没有用 callback，而是返回一个 EventEmitter 对象，调用者需要监听这个对象的 error 事件。这种方式在两种情况下很有用。一种情况是当在做一个可能会产生多个错误或多个结果的复杂操作的时候。例如，有一个请求一边从数据库取数据，一边把数据发送回客户端，而不是等待所有的结果一起到达。在这个例子里，没有用 callback，而是返回了一个 EventEmitter，每个结果会触发一个 row 事件，当所有结果发送完毕后会触发 end 事件，出现错误时会触发一个 error 事件。另一种情况是用在那些具有复杂状态机的对象上，这些对象往往伴随着大量的异步事件。例如，一个套接字是一个 EventEmitter，它可能会触发 connect、end、timeout、drain、close 等事件。这时很自然地可以把 error 作为另一种可以被触发的事件。在这种情况下，清楚知道 error 以及其他事件何时被触发很重要，同时要清楚被触发的还有什么事件、触发的顺序以及套接字是否在结束的时候处于关闭状态。

在大多数情况下，我们会把 callback 和 EventEmitter 归到同一个"异步错误传递"。如果有传递异步错误的需要，通常只要用其中的一种而不是同时使用。

使用 throw、callback、EventEmitter 取决于两点：

- 该异常是操作失败还是系统 Bug。
- 该函数本身是同步的还是异步的。

通用的准则就是，既可以同步传递错误（抛出），又可以异步传递错误（通过传给一个回调函数或者触发 EventEmitter 的 error 事件），但是不用同时使用。以这种方式，用户处理异常的时候可以选择用回调函数还是用 try/catch，但是不需要两种都用。具体用哪一种取决于异常是怎么传递的，必须在函数文档里说明清楚。

在编写新函数的时候，要用文档清楚地记录函数预期的参数，包括它们的类型、是否有其他约束（例如必须传入有效的 IP 地址）、可能会发生的合理的操作失败（例如无法解析主机名、连接服务器失败、所有的服务器端错误）、错误是怎么传递给调用者的（同步还是异步，同步可以使用 throw，异步可以使用 callback 和 EventEmitter）。

7.3 实战演练异常-错误处理

前面介绍了一些异常处理的分类和概念，本节通过代码演练的方式让读者理解异常-错误处理。

7.3.1 同步代码的异常捕获处理

【示例 7-1】同步代码中的异常使用 try{}catch 结构即可捕获处理。

```
try {
  throw new Error('error');
} catch(e) {
  console.log('异常被捕获了，我现在还可以继续执行了');
  console.log(e);
}
```

然后执行命令行时，可以看到如图 7.1 所示的结果，也会打印后面的 console.log 信息。

图 7.1　try{}catch 异常捕获

7.3.2 异步代码的错误处理

【示例 7-2】在异步代码中，使用 try{}catch 结构捕获处理效果如何呢？

```
try {
  setTimeout(()=>{
    throw new Error('错误信息');
  })
} catch (e) {
  console.error('error is:', e.message);
}
```

执行结果如图 7.2 所示，可以看出没有捕获到异步错误。

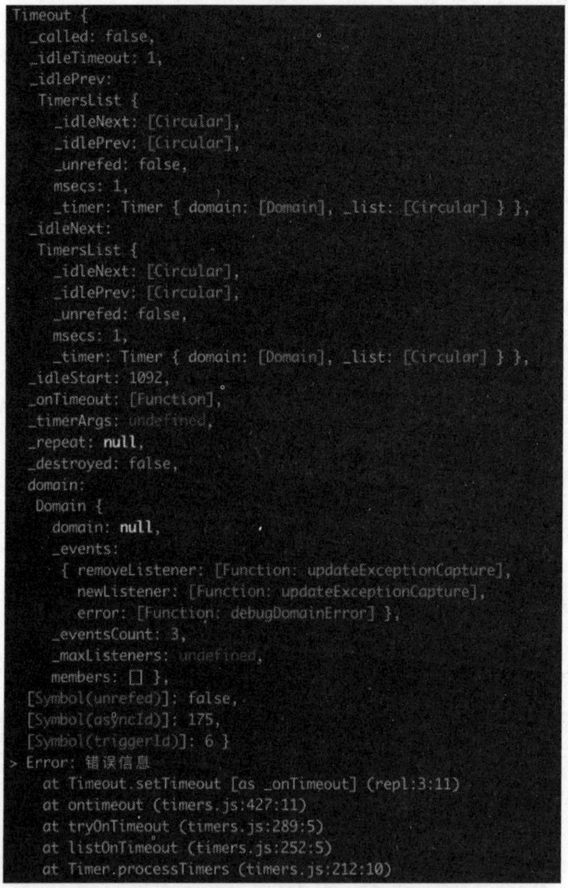

图 7.2 try{}catch 捕获不到异步错误

那么异步错误该怎么处理呢？首先换个思维，因为异常并不是事先准备好的，不能控制其到底在哪儿发生，所以以更高的角度，如监听应用进程的错误异常，从而捕获不能预料的错误异常，保证应用不至于崩溃。使用 process 的 uncaughtException 事件：

```
process.on('uncaughtException', (e)=>{
```

```
  console.error('process error is:', e.message);
});
```

以上代码从 process 上监听 uncaughtException 事件，可以捕获到整个进程包含异步中的错误信息，从而保证应用没有崩溃。

但是新的问题随之而来，因为异常不可预料地发生后，当异常出现时，直接从对应执行栈中断，而到 process 捕获的异常事件下，导致 V8 引擎的垃圾回收功能不能按照正常流程工作，然后开始出现内存泄漏问题。

相对于异常来说，内存泄漏也是一个不能忽视的严重问题，而 process.on('uncaughtException') 的做法很难保证不造成内存的泄漏。所以当捕获到异常时，显式地手动杀掉进程并重启 Node 进程，既可以保证释放内存，又保证了服务后续的正常可用。

```
process.on('uncaughtException', (e)=>{
  console.error('process error is:', e.message);
  process.exit(1);
  restartServer(); // 重启服务
});
```

7.3.3 使用 event 方式来处理异常

【示例 7-3】使用 event 方式来处理异常，代码如下：

```
const events = require('events');

// 创建一个事件监听对象
const emitter = new events.EventEmitter();

// 监听 error 事件
emitter.addListener('error', (e) => {
  // 处理异常信息
  console.log(11122222); // 能打印 11122222 说明异常捕获到了
  console.log(e);
});

// 触发 error 事件
emitter.emit('error', new Error('你代码出错了'));
```

运行效果如图 7.3 所示。

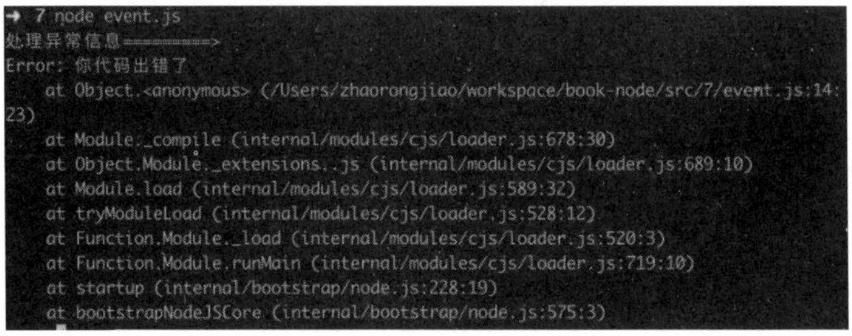

图 7.3 event 捕获异常

7.3.4 Callback 方式

【示例 7-4】读取一个文件，或者创建一个目录，测试代码如下：

```
const fs = require('fs');

fs.mkdir('/dir', (e) => {
  if (e) {
    console.log('异常信息处理');
    console.log(e);
  } else {
    console.log('创建目录成功');
  }
});
```

执行结果如图 7.4 所示。

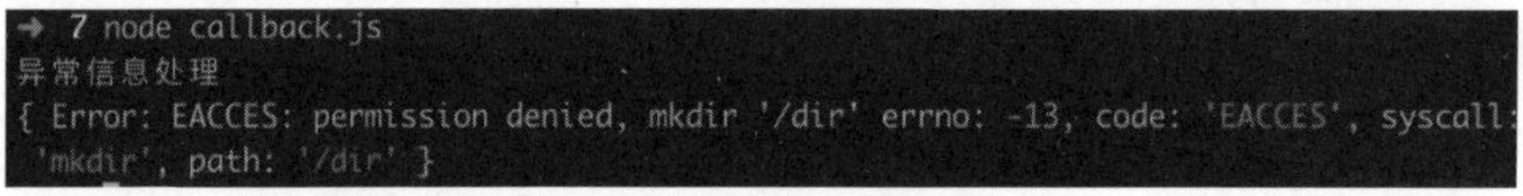

```
→ 7 node callback.js
异常信息处理
{ Error: EACCES: permission denied, mkdir '/dir' errno: -13, code: 'EACCES', syscall:
 'mkdir', path: '/dir' }
```

图 7.4　callback 捕获异常

7.3.5 Promise 方式

【示例 7-5】Promise 方式的示例代码如下：

```
new Promise((resolve, reject) => {
  throw new Error('error');
}).then(() => {
  // 一些逻辑代码
}).catch((e) => {
  console.log('能进来说明可以处理异常信息了');
  console.log(e);
});
```

执行结果如图 7.5 所示。

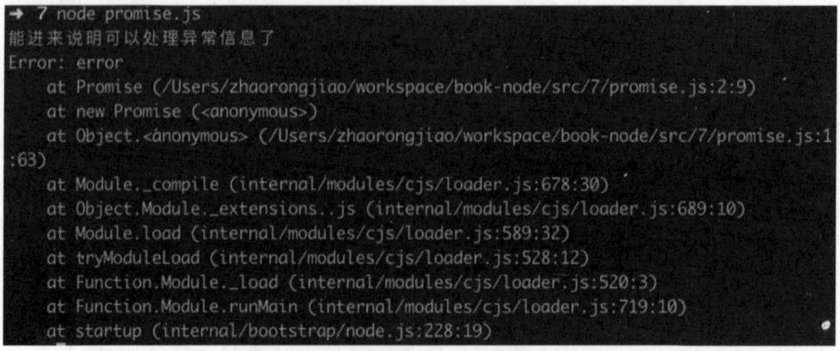

```
→ 7 node promise.js
能进来说明可以处理异常信息了
Error: error
    at Promise (/Users/zhaorongjiao/workspace/book-node/src/7/promise.js:2:9)
    at new Promise (<anonymous>)
    at Object.<anonymous> (/Users/zhaorongjiao/workspace/book-node/src/7/promise.js:1
:63)
    at Module._compile (internal/modules/cjs/loader.js:678:30)
    at Object.Module._extensions..js (internal/modules/cjs/loader.js:689:10)
    at Module.load (internal/modules/cjs/loader.js:589:32)
    at tryModuleLoad (internal/modules/cjs/loader.js:528:12)
    at Function.Module._load (internal/modules/cjs/loader.js:520:3)
    at Function.Module.runMain (internal/modules/cjs/loader.js:719:10)
    at startup (internal/bootstrap/node.js:228:19)
```

图 7.5　Promise 方式捕获异常

以上是处理同步代码，但是如果是异步代码呢？继续如下代码测试：

```
new Promise((resolve, reject) => {
  setTimeout(() => {
    throw new Error('error');
  }, 100);
}).then(() => {
  //一些逻辑代码
}).catch((e) => {
  console.log('能进来说明可以处理异常信息了');
  console.log(e);
});
```

执行结果如图 7.6 所示。可以看到，Promise 同样无法捕获异步代码中的异常信息。

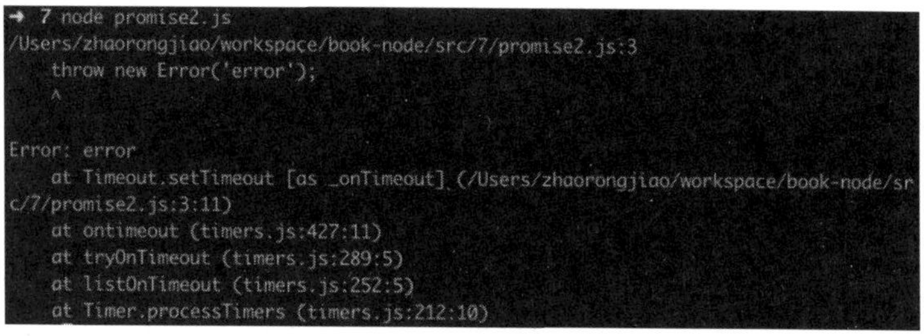

图 7.6　异步代码使用 Promise 捕获异常

Async/Await 也是基于 Promise 的，Promise 是无法捕获异步异常的，因此 Async/Await 也是没有办法捕获的。先看同步代码可以捕获到的异常，代码如下：

```
const testFunc = function() {
  return new Promise((resolve, reject) => {
    throw new Error('error');
  });
};

async function testAsync() {
  try {
    await testFunc();
  } catch (e) {
    console.log('能进来，说明异常能处理');
    console.log(e);
  }
}

testAsync();
```

执行结果如图 7.7 所示。

```
➜ 7 node async.js
能进来，说明异常能处理
Error: error
    at Promise (/Users/zhaorongjiao/workspace/book-node/src/7/async.js:3:11)
    at new Promise (<anonymous>)
    at testFunc (/Users/zhaorongjiao/workspace/book-node/src/7/async.js:2:10)
    at testAsync (/Users/zhaorongjiao/workspace/book-node/src/7/async.js:9:11)
    at Object.<anonymous> (/Users/zhaorongjiao/workspace/book-node/src/7/async.js:16:
1)
    at Module._compile (internal/modules/cjs/loader.js:678:30)
    at Object.Module._extensions..js (internal/modules/cjs/loader.js:689:10)
    at Module.load (internal/modules/cjs/loader.js:589:32)
    at tryModuleLoad (internal/modules/cjs/loader.js:528:12)
    at Function.Module._load (internal/modules/cjs/loader.js:520:3)
```

图 7.7　Async/Await 方式捕获同步异常

再看异步代码，如下所示：

```
const testFunc = function() {
  setTimeout(() => {
    console.log(1111);
    return new Promise((resolve, reject) => {
      throw new Error('error');
    });
  }, 100);
};

async function testAsync() {
  try {
    await testFunc();
  } catch (e) {
    console.log('能进来，说明异常能处理');
    console.log(e);
  }
}

testAsync();
```

执行结果如图 7.8 所示。

```
➜ 7 node async2.js
1111
(node:79878) UnhandledPromiseRejectionWarning: Error: error
    at Promise (/Users/zhaorongjiao/workspace/book-node/src/7/async2.js:5:13)
    at new Promise (<anonymous>)
    at Timeout.setTimeout [as _onTimeout] (/Users/zhaorongjiao/workspace/book-node/sr
c/7/async2.js:4:12)
    at ontimeout (timers.js:427:11)
    at tryOnTimeout (timers.js:289:5)
    at listOnTimeout (timers.js:252:5)
    at Timer.processTimers (timers.js:212:10)
(node:79878) UnhandledPromiseRejectionWarning: Unhandled promise rejection. This erro
r originated either by throwing inside of an async function without a catch block, or
 by rejecting a promise which was not handled with .catch(). (rejection id: 1)
(node:79878) [DEP0018] DeprecationWarning: Unhandled promise rejections are deprecate
d. In the future, promise rejections that are not handled will terminate the Node.js
process with a non-zero exit code.
```

图 7.8　Async/Await 方式捕获异步异常

7.3.6 使用 domain 模块

domain 模块把处理多个不同 IO 的操作作为一个组。注册事件和回调到 domain，当发生一个错误事件或抛出一个错误时，domain 对象会被通知，不会丢失上下文环境，也不会导致程序错误立即退出，与 process.on('uncaughtException')不同。

domain 模块可分为隐式绑定和显式绑定：

- 隐式绑定：把在 domain 上下文中定义的变量自动绑定到 domain 对象。
- 显式绑定：把不是在 domain 上下文中定义的变量以代码的方式绑定到 domain 对象。

【示例 7-6】示例代码如下：

```
const domain = require('domain');
const d = domain.create();

d.on('error', (err) => {
  console.log('err', err.message);
  console.log(needSend.message);
});

const needSend = { message: '需要传递给错误处理的一些信息' };
d.add(needSend);

function excute() {
  try {
    setTimeout(()=>{
      throw new Error('错误信息');
    });
  } catch (e) {
  console.error('error is:', e.message);
  }
};

d.run(excute);
```

domain 明显的优点是能把出问题时的一些信息传递给错误处理函数，可以做一些打点上报等处理工作，起码保证重启后的服务，程序员们知道发生了什么，有线索可查，也可以选择传递上下文进去，做一些后续处理。比如当服务出错的时候，可以把用户请求栈信息传给下游，返回告知用户服务异常，而不是用户一直等到请求自动超时。

```
...
d.add(res);
...
d.on('error', (err) => {
  console.log('err', err.message);
  res.end('服务器发生异常，请稍后再试！');
});
```

但是它和 process.on('uncaughtException')的做法一样，很难保证不造成内存的泄漏。

7.3.7 多进程模式加异常捕获后重启

上面的方式没有完美地解决问题，思考一下如何能够让异常发生后不崩溃，捕获异常后不造成内存泄漏，而且重启释放缓存不造成服务不可用呢？

一种比较好的方案是，以多进程（Cluster）的模式去部署应用，当某一个进程被异常捕获后，可以做一下打点上报后，开始重启释放内存，此时其他请求被接受后，其他进程依旧可以对外提供服务，当然前提是应用不能异常多的数都数不清。

【示例 7-7】下面将 cluster 和 domain 结合起来使用，以多进程的方式保证服务可用，同时可以将错误信息传递下去进行上报，并且保留错误出现的上下文环境，给用户返回请求，不让用户请求超时，再手动杀死异常进程，然后重启。

```
const cluster = require('cluster');
const os = require('os');
const http = require('http');
const domain = require('domain');

const d = domain.create();

if (cluster.isMaster) {
  const cpuNum = os.cpus().length;
  for (let i = 0; i < cpuNum; ++i) {
    cluster.fork()
  };
  // fork work log
  cluster.on('fork', worker=>{
    console.info(`${new Date()} worker${worker.process.pid}进程启动成功`);
  });
  // 监听异常退出进程，并重新 fork
  cluster.on('exit',(worker,code,signal)=>{
    console.info(`${new Date()} worker${worker.process.pid}进程启动异常退出`);
    cluster.fork();
  })
} else {
  http.createServer((req, res)=>{
    d.add(res);
    d.on('error', (err) => {
      console.log('记录的 err 信息', err.message);
      console.log('出错的 work id:', process.pid);
      // uploadError(err)   // 上报错误信息至监控
      res.end('服务器异常，请稍后再试');
      // 将异常子进程杀死
      cluster.worker.kill(process.pid);
    });
    d.run(handle.bind(null, req, res));
  }).listen(8080);
}

function handle(req, res) {
  if (process.pid % 2 === 0) {
```

```
    throw new Error('出错了');
  }
  res.end('response by worker: ${process.pid}');
};
```

运行以上代码：

```
node index.js
```

并运行：

```
curl 127.0.0.1:8080
```

结果如图 7.9 所示。

异常捕获运行结果如图 7.10 所示。

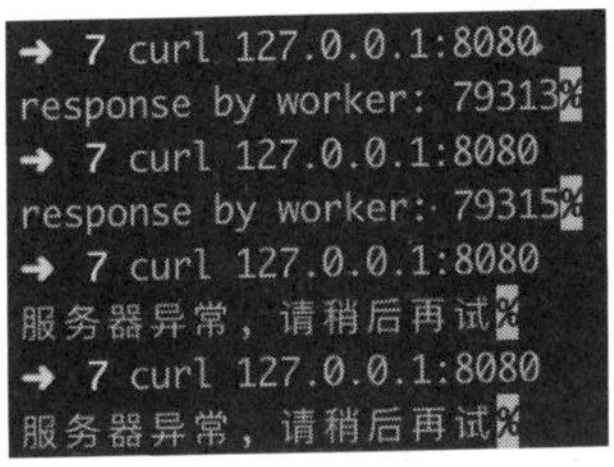

图 7.9　运行示例

图 7.10　异常捕获

第 8 章
◂Node.js的测试▸

代码部署之前，进行单元测试是十分必要的，这样能够有效并且持续保证代码质量。实践表明，高质量的单元测试还可以帮助我们完善自己的代码。本章将通过一些简单的测试案例介绍几款 Node.js 测试模块：Mocha、Should 和 SuperTest。

8.1 什么是单元测试

单元测试在项目中有着举足轻重的作用。在计算机编程中，单元测试（又称为模块测试，Unit Testing）是针对程序模块进行正确性检验的测试工作。程序单元是应用的最小可测试部件。在过程化编程中，一个单元就是单个程序、函数、过程等；对于面向对象编程，最小单元就是方法，包括基类（超类）、抽象类或者派生类（子类）中的方法。

JavaScript 是面向对象编程的，很多时候我们都需要将一个功能抽象成一个组件，方便团队其他开发者调用，那么我们理应保证给出的组件是正确可用的。在很长的一段时间里，前端都忽略了单元测试，或者说对于前端这种 GUI 编程来说，单元测试确实比较麻烦。随着 Node.js 的异军突起，针对 JavaScript 的单元测试框架如雨后春笋般出现，前端也逐渐玩起了单元测试。

单元测试的重要性是不言而喻的，经常存在的一个误区是：

- 单元测试不是测试人员的事情吗？
- 自己为什么要测试自己的代码？
- 单元测试的成本这么高，对于产品开发有意义吗？
- 我的代码足够健壮，不需要单元测试！

首先，认为测试是测试人员的事情是不负责的，测试人员应该更多地针对整体功能，而每一位工程师应该保证自己代码的准确性；其次，自己测试自己的代码更多时候是为了提高效率，如果写好了一个接口，没有经过测试就直接交给了别人，那么出错之后就需要接着调试，其中所花费的沟通成本会远大于编码成本；最后，在 GitHub 上面，所有 star 数量过万的 repo 都应该有自己的测试，所有的代码都必须经过测试才可投入使用。

单元测试根据主流的分类可以分成两类，分别是 BDD（Behavior-Driven Development，行为驱动开发和 TDD（Test-Driven Development，测试驱动开发）。TDD 的原理是在开发功能代码之前，先编写单元测试用例代码,测试代码确定需要编写什么产品代码。TDD 是 XP（Extreme Programming）的核心实践。测试驱动开发的流程是：

（1）开发人员写了一些测试代码。

（2）开发人员运行了这些测试用例，然后毫无疑问地，这些测试用例失败了，因为测试中提到的类和方法并没有实现。

（3）开发人员开始实现测试用例里面提到的方法。

（4）如果开发人员写好了某个功能点，就会欣喜地发现之前相对应的测试用例通过了。

（5）开发人员可以重构代码，并添加注释，完成后期工作。

BDD 的重点是通过与利益相关者的讨论取得对预期的软件行为的清醒认识。它通过用自然语言书写非程序员可读的测试用例扩展了测试驱动开发方法。行为驱动开发人员使用混合了领域中统一的语言的母语语言来描述他们的代码的目的。BDD 与 TDD 的主要区别是在写测试案例时的措辞，BDD 的测试案例更像是一份说明书，在详细描述软件的每一个功能。

8.2 一个简单的单元测试

所谓单元测试，就是对某个函数或者 API 进行正确性验证。

【示例 8-1】来看一个简单的示例：

```
function add(a, b)
{
    return a + b;
}
```

这是一个加法函数。执行一下：

```
> add = function(a, b){return a + b}
[Function: add]
> add(4)
NaN
```

当 add 函数仅给定一个参数 4 的时候，a 为 4，b 为 undefined，两者相加为 NaN。还有其他多种情况需要考虑：

- 你考虑过只有一个参数的场景吗？
- 给定一个参数时，NaN 是你想要的结果吗？
- 如果参数不是整数怎么办？

这时，就需要单元测试来验证各种可能的场景了。如果把 add 函数定义为两个整数相加，而其他输入返回 undefined，那么代码应该修改如下：

```
function add(a, b)
{
    if (typeof a === "number" && typeof b === "number")
    {
        return a + b;
    }
    else
    {
        return undefined;
    }
}
```

我们写代码的时候很容易陷入思维漏洞，而写测试的时候往往会考虑各种情况，这就是所谓的 TDD，因此，进行一定的单元测试是十分必要的：

- 验证代码的正确性。
- 避免修改代码时出错。
- 避免其他团队成员修改代码时出错。
- 便于自动化测试与部署。

8.3 Mocha

Mocha 是一个基于 Node.js 和浏览器的集合各种特性的 JavaScript 测试框架，并且可以让异步测试变得简单和有趣。Mocha 的测试是连续的，在正确的测试条件中遇到未捕获的异常时，会给出灵活且准确的报告。

安装使用：

```
npm install -g mocha
```

下面使用测试框架 Mocha 以及 Node.js 自带的断言库 Assert 编写一个简单的单元测试。

【示例 8-2】首先编写待测函数 calc.js：

```
exports.add = function(i, j) {
    return i + j;
};

exports.mul = function(i, j) {
    return i * j;
};
```

然后编写测试代码 demo.js：

```javascript
var assert = require('assert');

var calc = require('./calc.js');

describe('Calculator Tests', function () {
  describe('Addition Tests', function () {
    it('returns 1 + 1 = 2', function (done) {
      assert.equal(calc.add(1, 1), 2);
      done();
    });

    it('should returns 1 + -1 = 0', function (done) {
      assert.equal(calc.add(1, -1), 0);
      done();
    });
  });

  describe('Multiplication Tests', function () {
    it('returns 2 * 2 = 4', function (done) {
      assert.equal(calc.mul(2, 2), 4);
      done();
    });

    it('returns 0 * 4 = 4', function (done) {
      assert.equal(calc.mul(2, 2), 4);
      done();
    });
  });
});
```

测试代码中使用了测试框架 Mocha 提供的 it 函数，3 个 it 函数分别测试了 3 种不同的案例（Test Case）。it 函数的第 1 个参数为字符串，用于描述测试，一般会写期望得到的结果，例如"should return 3"；而第 2 个参数为函数，用于编写测试代码，一般是先调用被测试的函数或者 API，获取结果之后，使用断言库判断执行结果是否正确。

测试代码中使用了 Node.js 自带的断言库 Assert 的 assert.equal 函数，用于判定 add 函数返回的结果是否正确。assert.equal 成功时不会发生什么，而失败时会抛出一个 AssertionError。

使用 mocha 执行 test.js：

```
mocha test.js
```

下面为输出，表示测试案例全部通过：

```
Calculator Tests
    Addition Tests
```

```
    ✓ returns 1 + 1 = 2
    ✓ returns 1 + -1 = 0
  Multiplication Tests
    ✓ returns 2 * 2 = 4
    ✓ returns 0 * 4 = 4


 4 passing (7ms)
```

8.4 Assert

断言（Assertion）是单元测试中用来保证最小单元是否正常的检测方法。在程序设计中，断言是一种放在程序中的一阶逻辑（如一个结果为真或假的逻辑判断式），目的是为了标示与验证程序开发者预期的结果：当程序运行到断言的位置时，对应的断言应该为真。若断言不为真，则程序会中止运行，并给出错误消息。

断言用于检查程序在运行时是否满足期望。一旦断言检查失败，将会抛出异常停止整个应用的执行。在 Node.js 的 API 中定义了几种检测方法：

（1）assert.ok()判断结果是否为真：

```
assert.ok(false,'此为假');
//等同于
assert.equal(!!(add(1,2) === 3),true,'此为假');
```

（2）assert.ifError()判断实际值是否为一个假值（null,undefined,0,'',false），若实际值为真值，则抛出异常：

```
assert.ifError('');
assert.ifError(' ');  //Error: " "
```

（3）assert.equal()判断实际值与期望值是否相等：

```
assert.equal(add(1,2),3);
```

（4）assert.notEqual()判断实际值与期望值是否相等：

```
assert.notEqual(add(1,2),3);
```

（5）assert.deepStrictEqual()判断实际值与期望值是否深度相等：

```
assert.deepStrictEqual({a:1},{a:'1'});
```

（6）assert.notDeepStrictEqual 判断实际值与期望值是否不深度相等：

```
assert.notDeepStrictEqual({a:1},{a:'1'});
```

(7) assert.strictEqual() 判断实际值与期望值是否严格相等（相当于===）：

```
assert.strictEqual(1,true);
```

(8) assert.notStrictEqual() 判断实际值与期望值是否不严格相等（相当于！==）：

```
assert.notStrictEqual(1,true);
```

(9) assert.throws() 判断代码块是否抛出异常。assert.throws()语法形式如下：

```
assert.throws(fn[, error][, message])
```

其中，各参数形式为：

- fn <Function>
- error <RegExp> | <Function> | <Object> | <Error>
- message <string>

期望 fn 函数抛出异常。error 则用于指定抛出期望的错误。若指定 message，则当 fn 调用无法抛出错误或错误验证失败时，message 将附加到 AssertionError 提供的消息中。

```
//assert.throws(fn[, error][, message])
assert.throws(
    () => {
        throw new Error('错误值')
    },
    {
        message:'错误',
    }
)

assert.throws(
    () => {
        throw new Error('错误值')
    },
    {
        message:'错误',
    },'抛出异常不符合期望'
)
```

(10) assert.doesNotThrow() 判断代码块是否没有抛出异常。

assert.doesNotThrow()的语法如下：

```
assert.doesNotThrow(fn[, error][, message])
```

其中，各参数形式为：

- fn <Function>

- error <RegExp> | <Function>
- message <string>

断言 fn 函数不会抛出错误。若抛出错误且与 error 参数指定的类型相同，则抛出 AssertionError。若不同或 error 参数未定义，则错误将传播回调用方。

```
//assert.doesNotThrow(fn[, error][, message])
assert.doesNotThrow(
    () => {
        throw new TypeError('错误值');
    },
    /错误值/,
    '出错啦'
);

assert.doesNotThrow(
    () => {
        throw new TypeError('错误值');
    }
);
```

Node.js 自带的断言库 Assert 提供的函数有限，在实际工作中，Should 等第三方断言库更加强大和实用。

【示例 8-3】例如有 merge 函数：

```
function merge(a, b) {
  if (typeof a === "object" && typeof b === "object") {
    for (var property in b) {
      a[property] = b[property];
    }
    return a;
  } else {
    return undefined;
  }
}
```

编写测试代码如下：

```
require("should");
var merge = require("../merge.js");
// 当2个参数均为对象时
it("should success", function () {
  var a = {
    name: "HelloWorld",
    type: "SaaS"
  };
```

```
    var b = {
      service: "Real time bug monitoring",
      product: {
        frontend: "JavaScript",
        backend: "Node.js",
        mobile: "微信小程序"
      }
    };
    var c = merge(a, b);
    c.should.have.property("name", "HelloWorld");
    c.should.have.propertyByPath("product", "frontend").equal("JavaScript");
  });
  // 当只有1个参数时
  it("should return undefined", function () {
    var a = {
      name: "HelloWorld",
      type: "SaaS"
    };
    var c = merge(a);
    (typeof c).should.equal("undefined");
  });
```

测试代码中有三处使用 should：

```
c.should.have.property("name", "HelloWorld");
c.should.have.propertyByPath("product", "frontend").equal("JavaScript");
(typeof c).should.equal("undefined");
```

should 可以验证对象是否存在某属性，并验证其取值；还可以验证对象是否存在某个嵌套属性，并使用链式方式验证其取值。那么 should 为什么不能直接验证 c 的取值为 undefined 呢？例如：

```
c.should.equal(undefined);  // 这样写是错误的
```

should 会为每个对象添加 should 属性，然后通过该属性提供各种断言函数，我们可以使用这些函数验证对象的取值。对于 undefined，should 无法为其添加属性，因此失败。

通过 node 验证发现，导入 should 之后，空对象 a 增加了一个 should 属性。

```
> a = {}
> typeof a.should
'undefined'
> require("should")
> typeof a.should
'object'
```

8.5 测试 HTTP 接口

Node.js 经常用于编写 HTTP 接口，为前端提供服务。那么，如何编写 Node.js 单元测试对 HTTP 接口进行测试呢？

【示例 8-4】

（1）首先，编写一个简单的 HTTP 接口，代码示例如下：

```
var http = require("http");
var server = http.createServer((req, res) => {
  res.writeHead(200, {
    "Content-Type": "text/plain"
  });
  res.end("Hello LiLei");
});
server.listen(8000);
```

（2）使用 Mocha 访问接口并获取返回数据，对返回结果进行验证：

```
require("../server.js");
var http = require("http");
var assert = require("assert");
it("should return hello LiLei", function (done) {
  http.get("http://localhost:8000", function (res) {
    res.setEncoding("utf8");
    res.on("data", function (text) {
      assert.equal(res.statusCode, 200);
      assert.equal(text, "Hello LiLei");
      done();
    });
  });
});
```

> **注 意**
>
> http.get 访问 HTTP 接口是一个异步操作。Mocha 在测试异步代码时需要为 it 函数添加回调函数 done，在断言结束的地方调用 done，这样 Mocha 才能知道什么时候结束这个测试。

（3）使用 SuperTest 进行测试：

```
var request = require("supertest");
var server = require("../server.js");
var assert = require("assert");
it("should return hello LiLei", function (done) {
```

```
  request(server)
    .get("/")
    .expect(200)
    .expect(function (res) {
      assert.equal(res.text, "Hello LiLei");
    })
    .end(done);
});
```

SuperTest 封装了发送 HTTP 请求的接口,并且提供了简单的 expect 断言来判定接口返回结果。

8.6 代码覆盖率工具 istanbul

在测试代码时,有时关心是否所有代码都测试到位。这个指标就叫作"代码覆盖率"(Code Coverage)。它有 4 个测量维度:

- 行覆盖率(Line Coverage):是否每一行都执行。
- 函数覆盖率(Function Coverage):是否每个函数都调用。
- 分支覆盖率(Branch Coverage):是否每个 if 代码块都执行。
- 语句覆盖率(Statement Coverage):是否每个语句都执行。

istanbul 是 JavaScript 程序的代码覆盖率工具,本节介绍使用 istanbul 测试代码覆盖率。
安装 istanbul:

```
npm install -g istanbul
```

【示例 8-5】编写样例代码:

```
var a = 1;
var b = 1;
if ((a + b) > 2) {
  console.log('more than two');
}
```

使用 istanbul cover 命令就能得到覆盖率,如图 8.1 所示。

```
istanbul cover simple.js
```

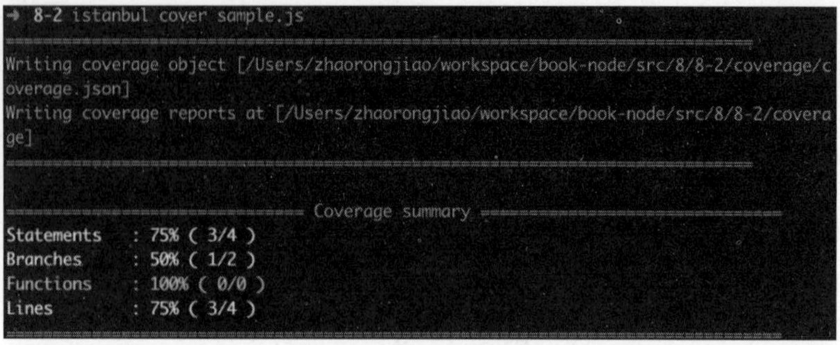

图 8.1　istanbul 示例

返回结果显示，simple.js 有 4 个语句（Statement），执行了 3 个；有两个分支（Branch），执行了 1 个；有 0 个函数，调用了 0 个；有 4 行代码，执行了 3 行。同时，还生成了一个 coverage 子目录，其中的 coverage.json 文件包含覆盖率的原始数据，coverage/lcov-report 是可以在浏览器中打开的覆盖率报告，其中有详细信息显示到底哪些代码没有覆盖到。

第 9 章
Node.js的数据处理

本章介绍如何使用 Node.js 来连接 MySQL 数据库、MongoDB 数据库、Redis 数据库，并对相应的数据库执行连接、增删改查等相关操作。

9.1 MySQL

MySQL 目前是使用量比较大的数据库，大中小型公司都在使用，所以本节的技术属于数据库操作的基础技术。

9.1.1 Node.js 连接 MySQL

要在 Node.js 中使用 MySQL，首先必须安装 mysql 模块依赖：

```
npm install mysql --save
```

然后在代码中引入依赖：

```
//引入数据库
var mysql=require('mysql');

//实现本地链接
var connection = mysql.createConnection({
   host: 'localhost',
   user: 'yf',
   password: '123456',
   database: 'yf'
})
```

数据库连接参数说明见表 9.1。

表 9.1 数据库连接参数说明

参数	描述
host	主机地址 （默认：localhost）
user	用户名

(续表)

参　数	描　　述
password	密码
port	端口号（默认：3306）
database	数据库名
charset	连接字符集（默认：'UTF8_GENERAL_CI'，注意字符集的字母都要大写）
localAddress	此 IP 用于 TCP 连接（可选）
socketPath	连接到 UNIX 域路径，当使用 host 和 port 时会被忽略
timezone	时区（默认：'local'）
connectTimeout	连接超时（默认：不限制；单位：毫秒）
stringifyObjects	是否序列化对象
typeCast	是否将列值转化为本地 JavaScript 类型值（默认：true）
queryFormat	自定义 query 语句格式化方法
supportBigNumbers	数据库支持 bigint 或 decimal 类型列时，需要设此 option 为 true（默认：false）
bigNumberStrings	supportBigNumbers 和 bigNumberStrings 启用强制 bigint 或 decimal 列以 JavaScript 字符串类型返回（默认：false）
dateStrings	强制 timestamp、datetime、data 类型以字符串类型返回，而不是 JavaScript Date 类型（默认：false）
debug	开启调试（默认：false）
multipleStatements	是否允许一个 query 中有多个 MySQL 语句（默认：false）
flags	用于修改连接标志
ssl	使用 ssl 参数（与 crypto.createCredenitals 参数格式一致）或一个包含 ssl 配置文件名称的字符串，目前只捆绑 Amazon RDS 的配置文件

9.1.2　数据库操作

本小节通过创建、查询、关闭等操作来演示 MySQL 数据库的常见应用。

1．创建数据库

【示例 9-1】例如，创建一个表 node_user，包含 3 列：id、name、age，并插入 5 条数据记录：

```
Source Database       : my_news_test
SET FOREIGN_KEY_CHECKS=0;

-- ----------------------------
-- Table structure for node_user
-- ----------------------------
DROP TABLE IF EXISTS `node_user`;
CREATE TABLE `node_user` (
  `id` int(11) NOT NULL AUTO_INCREMENT,
  `name` varchar(30) DEFAULT NULL,
```

```
  `age` int(8) DEFAULT NULL,
  PRIMARY KEY (`id`)
) ENGINE=InnoDB AUTO_INCREMENT=6 DEFAULT CHARSET=utf8;

-- ----------------------------
-- Records of node_user
-- ----------------------------
INSERT INTO `node_user` VALUES ('1', 'admin', '32');
INSERT INTO `node_user` VALUES ('2', 'dans88', '45');
INSERT INTO `node_user` VALUES ('3', '张三', '35');
INSERT INTO `node_user` VALUES ('4', 'ABCDEF', '88');
INSERT INTO `node_user` VALUES ('5', '李小二', '65');
```

2. 查询数据

【示例9-2】查询所有用户的语句是 SELECT * FROM node_user，使用 connection.query 方法执行对应的查询语句：

```
//引入数据库
var mysql=require('mysql');

//实现本地连接
var connection = mysql.createConnection({
    host: 'localhost',
    user: 'yf',
    password: '123456',
    database: 'yf'
})
// 定义查询方法
// 查询所有用户
var showuser="SELECT * FROM node_user";
// 删除名为张三的用户
var deleteuserSql="DELETE FROM node_user WHERE name='张三'"

//调用查询方法
connection.query(userAddSql,userAddSql_Params,function(err,result){

    if(err) throw err;

    console.log('show result:',result);
    console.log('show result:',result.affectedRows);

});
```

3. 结束数据库连接

使用 connection.end 结束数据库连接：

```
connection.end();

//另一个结束方法
//connection.destory();
```

9.1.3 使用 Sequelize 操作数据库

直接使用 mysql 包提供的接口，我们编写的代码比较底层，ORM（Object-Relational Mapping）技术把关系数据库的表结构映射到对象上。本小节介绍使用 Node 的 ORM 框架 Sequelize 来操作数据库。我们读写的都是 JavaScript 对象，Sequelize 帮我们把对象变成数据库中的行。

【示例 9-3】例如，使用 Sequelize 查询 pets 表，代码示例如下：

```
Pet.findAll()
  .then(function (pets) {
    for (let pet in pets) {
       console.log(`${pet.id}: ${pet.name}`);
    }
  }).catch(function (err) {
    // error
});
```

因为 Sequelize 返回的对象是 Promise，所以可以用 then() 和 catch() 分别异步响应成功和失败。也可以用 ES7 的 await 来调用任何一个 Promise 对象，代码示例如下：

```
var pets = await Pet.findAll();
```

await 只有一个限制，必须在 async 函数中调用。上面的代码直接运行需要修改如下：

```
(async () => {
   var pets = await Pet.findAll();
})();
```

1. 准备工作

在使用 Sequlize 操作数据库之前，我们先在 MySQL 中创建一个表来测试。在 test 数据库中创建一个 pets 表。test 数据库是 MySQL 安装后自动创建的用于测试的数据库。在 MySQL 客户端命令行中执行以下命令：

```
grant all privileges on test.* to 'www'@'%' identified by 'www';

use test;

create table pets (
```

```
    id varchar(50) not null,
    name varchar(100) not null,
    gender bool not null,
    birth varchar(10) not null,
    createdAt bigint not null,
    updatedAt bigint not null,
    version bigint not null,
    primary key (id)
) engine=innodb;
```

第一条 grant 命令是创建 MySQL 的用户名和口令，均为 www，并赋予操作 test 数据库的所有权限。第二条 use 命令把当前数据库切换为 test。第三条命令创建了 pets 表。

然后，我们根据前面的工程结构创建 hello-sequelize 工程，结构如下：

```
hello-sequelize/
|
+- .vscode/
|  |
|  +- launch.json <-- VSCode 配置文件
|
+- init.txt <-- 初始化 SQL 命令
|
+- config.js <-- MySQL 配置文件
|
+- app.js <-- 使用 koa 的 JS
|
+- package.json <-- 项目描述文件
|
+- node_modules/ <-- npm 安装的所有依赖包
```

然后，添加如下依赖包：

```
"sequelize": "3.24.1",
"mysql": "2.11.1"
```

> **注 意**
>
> mysql 是驱动，我们不直接使用。但是用到的 sequelize，需要使用 npm install 命令安装。

其中，config.js 实际上是一个简单的配置文件：

```
var config = {
    database: 'test',      // 使用哪个数据库
    username: 'www',       // 用户名
    password: 'www',       // 口令
    host: 'localhost',     // 主机名
    port: 3306             // 端口号，MySQL 默认3306
```

```
};

module.exports = config;
```

接下来，我们就可以在 app.js 中操作数据库了。

2. 用 Sequelize 建立对数据库的连接

使用 Sequelize 操作 MySQL 需要先做两个准备工作。

（1）创建一个 sequelize 对象实例：

```
const Sequelize = require('sequelize');
const config = require('./config');

var sequelize = new Sequelize(config.database, config.username,
config.password, {
    host: config.host,
    dialect: 'mysql',
    pool: {
        max: 5,
        min: 0,
        idle: 30000
    }
});
```

（2）定义模型 Pet，告诉 Sequelize 如何映射数据库表：

```
var Pet = sequelize.define('pet', {
    id: {
        type: Sequelize.STRING(50),
        primaryKey: true
    },
    name: Sequelize.STRING(100),
    gender: Sequelize.BOOLEAN,
    birth: Sequelize.STRING(10),
    createdAt: Sequelize.BIGINT,
    updatedAt: Sequelize.BIGINT,
    version: Sequelize.BIGINT
}, {
    timestamps: false
});
```

用 sequelize.define() 定义 Model 时，传入名称 pet，默认的表名就是 pets。第二个参数指定列名和数据类型，如果是主键，就需要更详细地指定。第三个参数是额外的配置，我们传入 { timestamps: false } 是为了关闭 Sequelize 的自动添加 timestamp 的功能。

接下来往数据库中插入数据，可以用 Promise 的方式编写：

```
var now = Date.now();

Pet.create({
    id: 'g-' + now,
    name: 'Gaffey',
    gender: false,
    birth: '2007-07-07',
    createdAt: now,
    updatedAt: now,
    version: 0
}).then(function (p) {
    console.log('created.' + JSON.stringify(p));
}).catch(function (err) {
    console.log('failed: ' + err);
});
```

也可以用 await 方式编写:

```
(async () => {
    var dog = await Pet.create({
        id: 'd-' + now,
        name: 'Odie',
        gender: false,
        birth: '2008-08-08',
        createdAt: now,
        updatedAt: now,
        version: 0
    });
    console.log('created: ' + JSON.stringify(dog));
})();
```

用 Uri 的方式进行连接:

```
// 使用 Uri 连接
var sequelize = new Sequelize('mysql://root:root@localhost:3306/webdb');
```

3. 验证连接

```
sequelize
    .authenticate()
    .then(() => {
        console.log('Connection has been established successfully.');
    })
    .catch(err => {
        console.error('Unable to connect to the database:', err);
    });
const Pet = sequelize.defint('pet', {
 id: {
```

```
        type: Sequelize.STRING(50),
        primaryKey: true
    },
    name: Sequelize.STRING(100),
    gender: Sequelize.BOOLEAN,
    birth: Sequelize.STRING(10),
    createdAt: Sequelize.BIGINT,
    updateAt: Sequelize.BIGINT,
    version: Sequelize.BIGINT
})
```

4. 查询数据

查询数据时，例如查询一条 name 值为 Gaffey 的数据，用 await 写法如下：

```
(async () => {
    var pets = await Pet.findAll({
        where: {
            name: 'Gaffey'
        }
    });
    console.log('find ${pets.length} pets:');
    for (let p of pets) {
        console.log(JSON.stringify(p));
    }
})();
```

为了使用复杂一些的查询，如模糊查询等，需要引入 Operator，例如：

```
// 引入 Operator
const Op = Sequelize.Op;


finddogbyid = async(id) => {

    var targetdogs = await Pet.findAll({
        where: {
            id: {
                [Op.like]: id + '%'
            }
        }
    });
    for (let o of targetdogs) {
        console.log('Target:' + JSON.stringify(o));
    }
    return targetdogs;
}
```

```
finddogbyid(15);
```

5. 增加数据

增加一条记录，示例如下：

```
var now = Date.now();

makedog = async() => {
   var dog = await Pet.create({
       id: now,
       name: 'Baibai',
       gender: false,
       birth: '2001-01-01',
       createdAt: now,
       updatedAt: now,
       version: 0
   });
   console.log('created:' + JSON.stringify(dog));
};
makedog();
```

6. 删除数据

将数据库中的记录删除，示例如下：

```
deletedog = async() => {
      var deldog = await Pet.findAll({
         where: {
            version: 0
         }
      });
      for (let o of deldog) {
         await o.destroy();
      }
   }
deletedog();
```

7. 修改数据

```
editdog = async() => {
      var needEditdogs = await Pet.findAll({
         where: {
            version: 0
         }
      });
      for (let needEditdog of needEditdogs) {
         needEditdog.gender = true;
```

```
            needEditdog.updatedAt = Date.now();
            needEditdog.version++;
            await needEditdog.save();
        }
    }
    editdog();
```

9.2 MongoDB

MongoDB 是一种文档导向数据库管理系统,由 C++编写而成,是一个基于分布式文件存储的数据库,旨在为 Web 应用提供可扩展的高性能数据存储解决方案。本节介绍如何使用 Node.js 来连接 MongoDB,并对数据库进行操作。

MongoDB 将数据存储为一个文档,数据结构由键值(key=>value)对组成。MongoDB 文档类似于 JSON 对象。字段值可以包含其他文档、数组及文档数组。

9.2.1 创建数据库

在 MongoDB 中创建一个数据库,首先创建一个 MongoClient 对象,然后配置好指定的 URL 和端口号。如果数据库不存在,MongoDB 就会创建数据库并建立连接。示例代码如下:

```
var MongoClient = require('mongodb').MongoClient;
var url = "mongodb://localhost:27017/books";

MongoClient.connect(url, { useNewUrlParser: true }, function(err, db) {
  if (err) throw err;
  console.log("数据库已创建!");
  db.close();
});
```

使用 createCollection() 方法来创建集合:

```
var MongoClient = require('mongodb').MongoClient;
var url = 'mongodb://localhost:27017/books';
MongoClient.connect(url, { useNewUrlParser: true }, function (err, db) {
   if (err) throw err;
   console.log('数据库已创建');
   var dbase = db.db("books");
   dbase.createCollection('site', function (err, res) {
      if (err) throw err;
      console.log("创建集合!");
      db.close();
   });
```

    });

9.2.2 数据库操作

1. 插入数据

（1）插入一条数据

【示例 9-4】以下实例我们连接数据库 books 的 site 表，并插入一条数据，使用 insertOne() 实现：

```
var MongoClient = require('mongodb').MongoClient;
var url = "mongodb://localhost:27017/";

MongoClient.connect(url, { useNewUrlParser: true }, function(err, db) {
   if (err) throw err;
   var dbo = db.db("books");
   var myobj = { name: "Node.js 实战", url: "www.books" };
   dbo.collection("site").insertOne(myobj, function(err, res) {
      if (err) throw err;
      console.log("文档插入成功");
      db.close();
   });
});
```

执行以下命令输出结果为：

```
$ node test.js
文档插入成功
```

从输出结果来看，数据已插入成功。也可以打开 MongoDB 的客户端查看数据，例如：

```
> show dbs
books   0.000GB         # 自动创建了 books 数据库
> show tables
site                    # 自动创建了 site 集合（数据表）
> db.site.find()
{ "_id" : ObjectId("5a794e36763eb821b24db854"), "name" : "Node.js 实战", "url" : "www.books" }
>
```

（2）插入多条数据

【示例 9-5】如果要插入多条数据，可以使用 insertMany()：

```
var MongoClient = require('mongodb').MongoClient;
var url = "mongodb://localhost:27017/";
```

```
MongoClient.connect(url, { useNewUrlParser: true }, function(err, db) {
    if (err) throw err;
    var dbo = db.db("books");
    var myobj = [
        { name: '书本集合 ', url: 'https://c.books.com', type: 'cn'},
        { name: 'Google', url: 'https://www.google.com', type: 'en'},
        { name: 'Facebook', url: 'https://www.google.com', type: 'en'}
      ];
    dbo.collection("site").insertMany(myobj, function(err, res) {
        if (err) throw err;
        console.log("插入的文档数量为: " + res.insertedCount);
        db.close();
    });
});
```

res.insertedCount 为插入的条数。

2. 查询数据

【示例 9-6】可以使用 find()来查找数据，find()可以返回匹配条件的所有数据。如果未指定条件，find()就返回集合中的所有数据。

```
var MongoClient = require('mongodb').MongoClient;
var url = "mongodb://localhost:27017/";

MongoClient.connect(url, { useNewUrlParser: true }, function(err, db) {
    if (err) throw err;
    var dbo = db.db("books");
    dbo.collection("site"). find({}).toArray(function(err, result) { // 返回集合中所有数据
        if (err) throw err;
        console.log(result);
        db.close();
    });
});
```

以下检索 name 为"Node.js 实战"的实例查询指定条件的数据：

```
var MongoClient = require('mongodb').MongoClient;
var url = "mongodb://localhost:27017/";

MongoClient.connect(url, { useNewUrlParser: true }, function(err, db) {
    if (err) throw err;
    var dbo = db.db("books");
     var whereStr = {"name":'Node.js 实战'};  // 查询条件
    dbo.collection("site").find(whereStr).toArray(function(err, result) {
        if (err) throw err;
```

```
        console.log(result);
        db.close();
    });
});
```

执行以下命令输出结果为：

```
[ { _id: 5a794e36763eb821b24db854,
    name: 'Node.js 实战',
    url: 'www.books' } ]
```

3. 更新数据

【示例 9-7】我们也可以对数据库的数据进行修改，以下实例将 name 为 "Node.js 实战" 的 URL 改为 "https://www.books.com"：

```
var MongoClient = require('mongodb').MongoClient;
var url = "mongodb://localhost:27017/";

MongoClient.connect(url, { useNewUrlParser: true }, function(err, db) {
    if (err) throw err;
    var dbo = db.db("books");
    var whereStr = {"name":'Node.js 实战'};  // 查询条件
    var updateStr = {$set: { "url" : "https://www.books.com" }};
    dbo.collection("site").updateOne(whereStr, updateStr, function(err, res) {
        if (err) throw err;
        console.log("文档更新成功");
        db.close();
    });
});
```

执行成功后，进入 MongoDB 管理工具查看数据已修改：

```
> db.site.find().pretty()
{
    "_id" : ObjectId("5a794e36763eb821b24db854"),
    "name" : "Node.js 实战",
    "url" : "https://www.books.com"     // 已修改为 https
}
```

如果要更新所有符合条件的文档数据，就可以使用 updateMany()：

```
var MongoClient = require('mongodb').MongoClient;
var url = "mongodb://localhost:27017/";

MongoClient.connect(url, { useNewUrlParser: true }, function(err, db) {
    if (err) throw err;
    var dbo = db.db("books");
```

```
        var whereStr = {"type":'en'};  // 查询条件
        var updateStr = {$set: { "url" : "https://www.books.com" }};
        dbo.collection("site").updateMany(whereStr, updateStr, function(err, res)
{
        if (err) throw err;
         console.log(res.result.nModified + " 条文档被更新");
        db.close();
    });
});
```

result.nModified 为更新的条数。

4. 删除数据

【示例 9-8】以下实例将 name 为 "Node.js 实战" 的数据删除：

```
var MongoClient = require('mongodb').MongoClient;
var url = "mongodb://localhost:27017/";

MongoClient.connect(url, { useNewUrlParser: true }, function(err, db) {
    if (err) throw err;
    var dbo = db.db("books");
    var whereStr = {"name":'Node.js 实战'};  // 查询条件
    dbo.collection("site").deleteOne(whereStr, function(err, obj) {
        if (err) throw err;
        console.log("文档删除成功");
        db.close();
    });
});
```

执行成功后，进入 MongoDB 管理工具确认数据已删除：

```
> db.site.find()
>
```

【示例 9-9】如果要删除多条语句，可以使用 deleteMany()方法：

```
var MongoClient = require('mongodb').MongoClient;
var url = "mongodb://localhost:27017/";

MongoClient.connect(url, { useNewUrlParser: true }, function(err, db) {
    if (err) throw err;
    var dbo = db.db("books");
    var whereStr = { type: "en" };  // 查询条件
    dbo.collection("site").deleteMany(whereStr, function(err, obj) {
        if (err) throw err;
        console.log(obj.result.n + " 条文档被删除");
        db.close();
```

```
      });
   });
```

obj.result.n 为删除的条数。

5. 排序

使用 sort()方法进行排序，该方法接收一个参数，规定是升序（1）还是降序（-1），例如：

```
{ type: 1 }           // 按 type 字段升序
{ type: -1 }          // 按 type 字段降序
```

【示例 9-10】排序：

```
var MongoClient = require('mongodb').MongoClient;
var url = "mongodb://localhost:27017/";

MongoClient.connect(url, { useNewUrlParser: true }, function(err, db) {
   if (err) throw err;
   var dbo = db.db("books");
   var mysort = { type: 1 };
   dbo.collection("site").find().sort(mysort).toArray(function(err, result) {
      if (err) throw err;
      console.log(result);
      db.close();
   });
});
```

6. 查询分页

【示例 9-11】如果要设置指定的返回条数，就可以使用 limit()方法，该方法只接收一个参数，用于指定返回的条数。limit(2)表示读取了两条数据：

```
var MongoClient = require('mongodb').MongoClient;
var url = "mongodb://localhost:27017/";

MongoClient.connect(url, { useNewUrlParser: true }, function(err, db) {
   if (err) throw err;
   var dbo = db.db("books");
   dbo.collection("site").find().limit(2).toArray(function(err, result) {
      if (err) throw err;
      console.log(result);
      db.close();
   });
});
```

如果要指定跳过的条数，就可以使用 skip()方法。例如跳过前面两条数据，读取两条数据：

```
var MongoClient = require('mongodb').MongoClient;
var url = "mongodb://localhost:27017/";

MongoClient.connect(url, { useNewUrlParser: true }, function(err, db) {
    if (err) throw err;
    var dbo = db.db("books");
    dbo.collection("site").find().skip(2).limit(2).toArray(function(err, result) {
        if (err) throw err;
        console.log(result);
        db.close();
    });
});
```

7. 连接操作

MongoDB 不是一个关系型数据库，但我们可以使用 $lookup 来实现左连接。例如有两个集合，数据分别为：

集合 1：orders

```
[
  { _id: 1, product_id: 154, status: 1 }
]
```

集合 2：products

```
[
  { _id: 154, name: '笔记本电脑' },
  { _id: 155, name: '耳机' },
  { _id: 156, name: '台式电脑' }
]
```

【示例 9-12】使用 $lookup 实现左连接：

```
var MongoClient = require('mongodb').MongoClient;
var url = "mongodb://127.0.0.1:27017/";

MongoClient.connect(url, { useNewUrlParser: true }, function(err, db) {
  if (err) throw err;
  var dbo = db.db("books");
  dbo.collection('orders').aggregate([
    { $lookup:
      {
        from: 'products',           // 右集合
        localField: 'product_id',   // 左集合 join 字段
        foreignField: '_id',        // 右集合 join 字段
        as: 'orderdetails'          // 新生成字段（类型 array）
```

```
      }
    }
  ]).toArray(function(err, res) {
    if (err) throw err;
    console.log(JSON.stringify(res));
    db.close();
  });
});
```

8. 删除集合

【示例 9-13】可以使用 drop()方法来删除集合：

```
var MongoClient = require('mongodb').MongoClient;
var url = "mongodb://localhost:27017/";

MongoClient.connect(url, { useNewUrlParser: true }, function(err, db) {
    if (err) throw err;
    var dbo = db.db("books");
    // 删除 test 集合
    dbo.collection("test").drop(function(err, delOK) {   // 执行成功 delOK 返回 true, 否则返回 false
        if (err) throw err;
        if (delOK) console.log("集合已删除");
        db.close();
    });
});
```

9.3 Redis

Redis（Remote Dictionary Server）是一个开源的使用 ANSI C 语言编写、遵守 BSD 协议、支持网络、可基于内存亦可持久化的日志型 Key-Value 数据库，并提供多种语言的 API。它通常被称为数据结构服务器，因为值（Value）可以是字符串（String）、哈希（Hash）、列表（List）、集合（Sets）和有序集合（Sorted Sets）等类型。

9.3.1 Node.js 连接 Redis

安装 redis 模块：

```
npm install redis -S
```

【示例 9-14】连接 redis 代码示例：

```
var redis = require('redis');
```

```
var client = redis.createClient(6379,'127.0.0.1');

client.auth(123456);  // 如果没有设置密码，那么不需要这一步
client.on('connect', function () {
    client.set('hello','This is a value');
})
```

连接后效果如图 9.1 所示。

```
127.0.0.1:6379> keys *
1) "hello"
```

图 9.1 redis 连接

【示例 9-15】获取 redis 中的值：

```
var redis = require('redis');

var client = redis.createClient(6379,'127.0.0.1');

client.on('connect', function () {
   // set 语法
   client.set('hello','This is a value');
   // get 语法
   client.get('hello',function (err,v) {
       console.log("redis get hello err,v",err,v);
   })
})
```

效果如图 9.2 所示。

```
127.0.0.1:6379> get hello
"This is a value"
```

图 9.2 获取 redis 中的值

把存储对象改成 JSON 对象：

```
var redis = require('redis');

var client = redis.createClient(6379,'127.0.0.1');
client.on('connect', function () {
   client.set('hello',{name:"jacky",age:22});

   client.get('hello',function (err,v) {
       console.log("redis get hello err,v",err,v);
   })
})
```

程序将会报错，因为 Redis 中存储的是字符串对象，如图 9.3 所示。

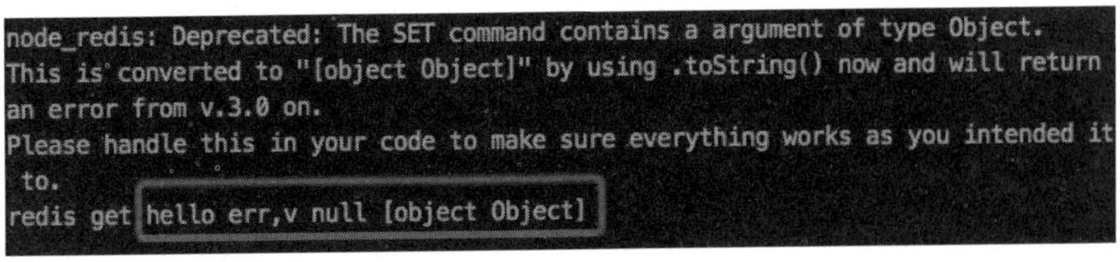

图 9.3　JSON 获取异常

重写 toString 方法：

```
Object.prototype.toString = function(){
    return JSON.stringify(this);
};
```

输出正常，如图 9.4 所示。

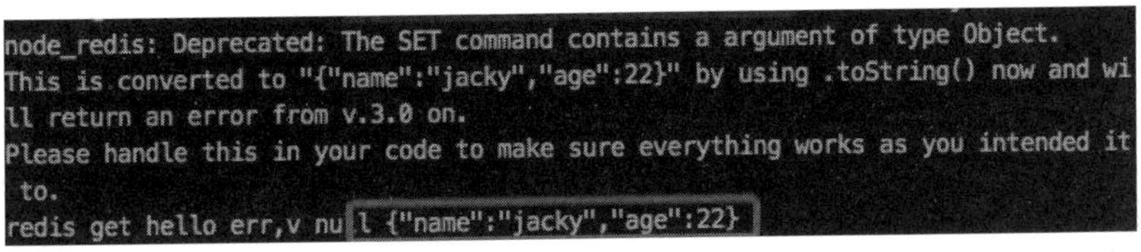

图 9.4　JSON 获取正确

9.3.2　列表——List

为了方便，把连接 Redis 的代码写成模块的方式。创建 client.js 文件，写入如下代码：

```
var redis = require('redis');

module.exports = redis.createClient(6379, 127.0.0.1);
```

【示例 9-16】操作 list 就是操作双向链表：

```
var client = require('./client');

//先清除数据
client.del('testLists');
client.rpush('testLists','a');
client.rpush('testLists','b');
client.rpush('testLists','c');
client.rpush('testLists','d');
client.rpush('testLists','e');
client.rpush('testLists','1');

client.lpop('testLists',function (err,v) {
    console.log('client.lpop , err , v : ' , err,v);
})
```

```
client.rpop('testLists',function (err,v) {
    console.log('client.rpop , err, v ',err,v);
})

client.lrange('testLists',0.,-1, function (err,lists) {
    console.log('client.lrange , err ,lists: ',err,lists);
})
```

运行结果如图 9.5 所示。

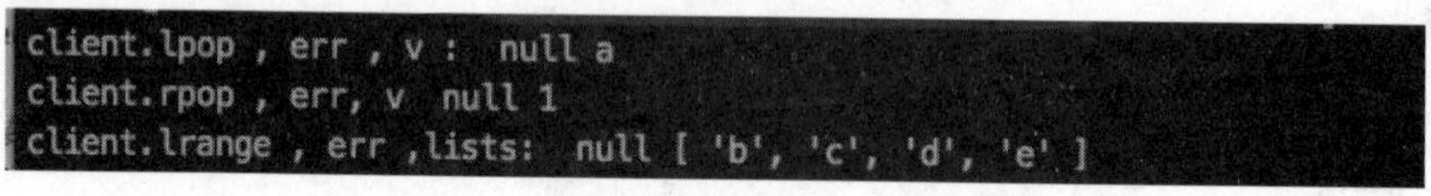

```
client.lpop , err, v :  null a
client.rpop , err, v  null 1
client.lrange , err ,lists:  null [ 'b', 'c', 'd', 'e' ]
```

图 9.5 列表操作

9.3.3 集合——Set

【示例 9-17】Set 是一种无序不可重复的集合，操作示例如下：

```
var client = require('./client');

client.sadd('testSet', 1);
client.sadd('testSet', 'a');
client.sadd('testSet', 'bb');

//不可重复
client.sadd('testSet', 'bb');

client.smembers('testSet', function(err, v){
    console.log('client.smembers err, v:', err, v);
});
```

运行结果如图 9.6 所示。

```
client.smembers err, v: null [ 'bb', 'a', '1' ]
```

图 9.6 集合操作

9.3.4 消息中介

【示例 9-18】首先编写订阅方，创建 sub.js 文件并输入如下代码：

```
var client = require('./client');

client.subscribe('testPublish');

client.on('message', function(channel, msg){
```

```
    console.log('client.on message, channel:', channel, ' message:', msg);
});
```

接着编写发布方代码,创建 pub.js 文件,并输入代码:

```
var client = require('./client');

client.publish('testPublish', 'message from pub.js');
```

先启动订阅方,然后启动发布方,便可看见效果,如图 9.7 所示。

```
client.on message, channel: testPublish  message: message from pub.js
```

图 9.7　发布/订阅

第 10 章 实战：使用原生JavaScript开发Node.js案例

本章介绍如何使用原生 JavaScript 开发基于 Node.js 的 Web 应用，包括如何启动 HTTP 服务器、如何处理请求路由和请求处理程序、如何使用非阻塞式方法、如何处理 POST 请求，最后介绍如何实现文件上传。

10.1 项目任务

使用原生 JavaScript 开发基于 Node.js 的 Web 应用，需要实现如下几个功能：

- 应用通过浏览器可访问。
- 实现文件上传表单和功能。
- 可预览文件上传的内容。

要实现如上几个功能，该项目需要提供 HTTP 服务、路由服务、请求处理服务、POST 支持以及预览服务：

- HTTP 服务：提供一个 Web 页面，因此需要一个 HTTP 服务器。
- 路由服务：对于不同的请求，根据请求的 URL，服务器需要给予不同的响应，因此需要一个路由，用于把请求对应到请求处理程序（Request Handler）。
- 请求处理：当请求被服务器接收并通过路由传递之后，需要对其进行处理，因此我们需要最终的请求处理程序。
- POST 支持：路由还应该能处理 POST 数据，并且把数据封装成更友好的格式传递给请求处理程序，因此需要请求数据处理功能。
- 预览服务：不仅要处理 URL 对应的请求，还要把内容显示出来，这意味着我们需要一些视图逻辑供请求处理程序使用，以便将内容发送给用户的浏览器。

用户需要上传图片，所以需要上传处理功能来处理这方面的细节。现在我们就来介绍如何实现一个 HTTP 服务。

10.2 HTTP 服务器

创建一个 server.js 文件，在 server.js 文件中编写如下示例代码：

```
const http = require("http");

http.createServer(function(request, response) {
  response.writeHead(200, {"Content-Type": "text/plain"});
  response.write("Hello World! ");
  response.end();
}).listen(8888);
```

上面的代码实现了一个完整的 Node.js 服务器。打开终端，在终端中输入命令 node server.js，在浏览器中打开 http://127.0.0.1:8888/，如图 10.1 所示。

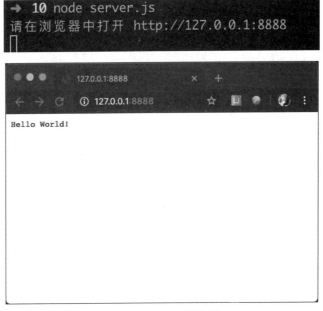

图 10.1　HTTP 服务器

如图 10.1 所示，一个基础的 HTTP 服务器开发完毕。本示例中，使用 Node.js 自带的 http 模块，并且把它赋值给 http 变量。

接下来，调用 http 模块提供的函数 createServer，createServer 函数返回一个对象，该对象含有 listen 方法，listen 方法有一个数值参数，指定这个 HTTP 服务器监听的端口号。

可以用这样的代码来启动服务器并侦听 8888 端口：

```
var http=require("http");

var server=http.createServer();
```

```
server.listen(8888);
```

这段代码只会启动一个侦听 8888 端口的服务器，它不做任何别的事情，不会应答请求，无法添加业务逻辑。因此，示例代码中使用函数传递让 HTTP 服务器工作，向 createServer 函数传递了一个匿名函数。或者改成箭头函数，如下示例代码也可以达到同样的目的：

```
//请求（require）Node.js 自带的 http 模块，并且把它赋值给 http 变量
let http = require("http");

//箭头函数
let onRequest = (request, response) => {
    response.writeHead(200, {"Content-Type": "text/plain"});
    response.write("Hello World");
    response.end();
}
//把函数当作参数传递
http.createServer(onRequest).listen(8888);

console.log("请在浏览器中打开 http://127.0.0.1:8888...");
```

事件驱动是 Node.js 原生的工作方式，当我们使用 http.createServer 方法的时候，当然不仅仅是侦听某个端口的服务器，还需要在服务器收到一个 HTTP 请求的时候添加具体的业务逻辑。

创建服务器，并且向创建它的方法传递了一个函数。无论何时，我们的服务器收到一个请求，这个函数就会被调用，即回调函数。我们给某个方法传递了一个函数，这个方法在有相应事件发生时，调用这个函数来进行回调。

示例代码如下：

```
//请求（require）Node.js 自带的 http 模块，并且把它赋值给 http 变量
let http = require("http");

//箭头函数
let onRequest = (request, response) => {
    console.log("Request received.");
    response.writeHead(200, {"Content-Type": "text/plain;charset=utf-8"});
    response.write("Hello, HTTP 服务器已启动");
    response.end();
}
//把函数当作参数传递
http.createServer(onRequest).listen(8888);

console.log("Server has started.");
console.log("请在浏览器中打开 http://127.0.0.1:8888");
```

运行效果如图 10.2 所示。

图 10.2　HTTP 服务器浏览器效果示例

接下来简单分析一下服务器代码中剩下的部分，也就是回调函数 onRequest()的主体部分。当回调启动，onRequest()函数被触发时，request 和 response 两个参数被传入。request 和 response 是对象，可以使用它们的方法来处理 HTTP 请求的细节，并且响应请求，例如向发出请求的浏览器发回响应内容。

当收到请求时，使用 response.writeHead()函数发送一个 HTTP 状态 200 和 HTTP 头的内容类型（Content-Type），使用 response.write()函数在 HTTP 响应主体中发送文本。最后调用 response.end()完成响应。目前尚未使用 request 对象。

10.3　服务端模块化

在 10.2 节的示例中，使用了 Node.js 自带的 http 模块，在代码中请求引入 http 模块并把返回值赋给一个本地变量，该本地变量变成了一个拥有所有 http 模块所提供的公共方法的对象。

类似地，可以自定义模块。将 server.js 文件的内容修改如下：

```
const http = require("http");

const start = () => {
 //箭头函数
 const onRequest = (request, response) => {
   console.log("Request received.");
   response.writeHead(200, {
     "Content-Type": "text/plain;charset=utf-8"
   });
```

```
    response.write("Hello World");
    response.end();
  }
  //把函数当作参数传递
  http.createServer(onRequest).listen(8888);

  console.log("Server has started.");
  console.log("请在浏览器中打开 http://127.0.0.1:8888");
}

//导出'server'对象，对象中包含一个start函数
//对象格式为
/**
 * {
 *     start
 * }
 */

//这个对象导入其他文件中即可使用，可以用任意的名字来接收这个对象

exports.start = start;
```

在 server.js 当前的文件路径下新建一个 index.js 文件。内容如下：

```
//从'server'模块中导入 server 对象

let server = require('./server');

//启动服务器
server.start();
```

运行 index.js 文件，控制台效果如图 10.3 所示。

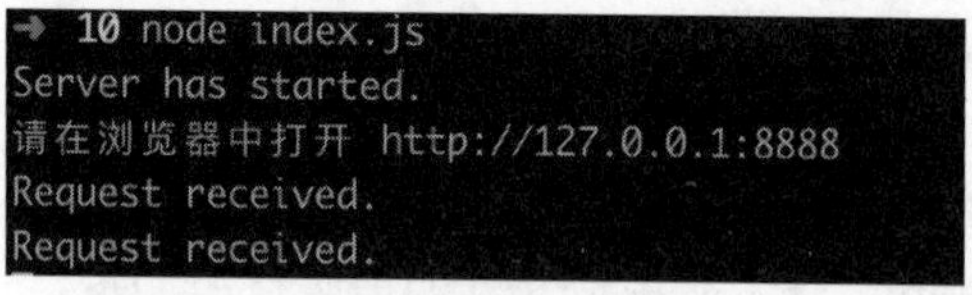

```
→ 10 node index.js
Server has started.
请在浏览器中打开 http://127.0.0.1:8888
Request received.
Request received.
```

图 10.3　模块化

10.4 设计请求路由

需要为路由提供请求的 URL 和其他所需的 GET 及 POST 参数，随后路由需要根据这些数据来执行相应的代码，如一系列在接收到请求时真正工作的处理程序。

因此，需要从 HTTP 请求中提取出请求的 URL 以及 GET/POST 参数。我们需要的所有数据都会包含在 request 对象中，该对象作为 onRequest()回调函数的第一个参数传递过去。但是为了解析这些数据，我们需要额外的 Node.js 模块，它们分别是 url 和 querystring 模块。也可以使用 querystring 模块来解析 POST 请求体中的参数。现在我们给 onRequest()函数加上一些逻辑，用来找出浏览器请求的 URL 路径：

```javascript
const http = require("http");
const url = require("url");

const start = () => {
  //箭头函数
  const onRequest = (request, response) => {
    const pathname = url.parse(request.url).pathname;
    console.log("Request received.");
    console.log("Request for " + pathname + " received.");

    response.writeHead(200, {
      "Content-Type": "text/plain;charset=utf-8"
    });
    response.write("Hello World! ");
    response.end();
  }
  //把函数当作参数传递
  http.createServer(onRequest).listen(8888);

  console.log("Server has started.");
  console.log("请在浏览器中打开 http://127.0.0.1:8888");
}

exports.start = start;
```

接下来，在终端执行 node index.js 命令，如下所示：

```
node index.js
Server has started.
请在浏览器中打开 http://127.0.0.1:8888
```

在浏览器中打开 http://127.0.0.1:8888，浏览器展示效果如图 10.4 所示。

图 10.4　设计路由

这时应用可以通过请求的 URL 路径来区别不同请求，使得代码中可以使用路由来将请求以 URL 路径为基准映射到处理程序上，这意味着在我们所要构建的应用中，来自/start 和/upload 的请求可以使用不同的代码来处理。稍后我们将看到这些内容是如何整合到一起的。

现在我们可以来编写路由了，建立一个名为 router.js 的文件，添加以下内容：

```javascript
function route(pathname) {
  console.log("About to route a request for " + pathname);
}

exports.route = route;
```

这段代码很简单，没有任何业务逻辑。在添加更多的逻辑之前，先来看看如何把路由和服务器整合起来。

首先，对服务器的 start() 函数进行扩展，以便将路由函数作为参数传递过去：

```javascript
let http = require("http");
let url = require("url");

//用一个函数将之前的内容包裹起来
let start = (route) => {
        //箭头函数
    let onRequest = (request, response) => {

        let pathname = url.parse(request.url).pathname;
        console.log("Request for " + pathname + " received.");
        route(pathname);

        response.writeHead(200,                            {"Content-Type": "text/plain;charset=utf-8"});
        response.write("Hello World!");
        response.end();
    }
    //把函数当作参数传递
    http.createServer(onRequest).listen(8888);

    console.log("Server has started.");
    console.log("请在浏览器中打开 http://127.0.0.1:8888...");
}

exports.start = start;
```

同时，相应地扩展 index.js，使得路由函数可以被注入服务器中：

```javascript
//从'server'模块中导入 server 对象

let server = require('./server');
```

```
let router = require("./router");

//启动服务器
server.start(router.route);
```

现在使用 node index.js 命令启动应用，随后在浏览器中请求 http://127.0.0.1:8888，你将会看到应用输出相应的信息，这表明我们的 HTTP 服务器已经在使用路由模块了，并会将请求的路径传递给路由：

```
node index.js
Server has started.
请在浏览器中打开 http://127.0.0.1:8888...
Request for / received.
About to route a request for /
```

现在我们的 HTTP 服务器和请求路由模块已经如期望的那样可以相互调用了。当然，这还远远不够，路由是指需要针对不同的 URL 使用不同的处理逻辑，处理/start 的"业务逻辑"就应该和处理/upload 的不同。在现在的实现下，路由过程会在路由模块中"结束"，并且路由模块并不是真正针对请求"采取行动"的模块，否则当我们的应用程序变得更为复杂时，将无法很好地扩展。

创建请求处理程序，添加 requestHandlers 的模块，并对每一个请求处理程序添加一个占位用函数，随后将这些函数作为模块的方法导出：

```
function start() {
  console.log("Request handler 'start' was called.");
}

function upload() {
  console.log("Request handler 'upload' was called.");
}

exports.start = start;
exports.upload = upload;
```

现在将一系列请求处理程序通过一个对象来传递，并且需要使用松耦合的方式将这个对象注入 route()函数中。先将这个对象引入主文件 index.js 中：

```
//从'server'模块中导入 server 对象

let server = require('./server');
let router = require("./router");
let requestHandlers = require("./requestHandlers");

//对象构造
var handle = {}
```

```
handle["/"] = requestHandlers.start;
handle["/start"] = requestHandlers.start;
handle["/upload"] = requestHandlers.upload;

//启动服务器
server.start(router.route, handle);
```

正如所见，将不同的 URL 映射到相同的请求处理程序上，只要在对象中添加一个键为"/"的属性，对应 requestHandlers.start 即可，这样我们就可以干净简洁地配置/start 和/的请求，使之都交由 start 这一处理程序来处理。在完成了对象的定义后，我们把它作为额外的参数传递给服务器，为此将 server.js 修改如下：

```
const http = require("http");
const url = require("url");

//用一个函数将之前的内容包裹起来
const start = (route,handle) => {
    //箭头函数
    const onRequest = (request, response) => {

        const pathname = url.parse(request.url).pathname;
        console.log("Request for " + pathname + " received.");
        route(handle,pathname);

        response.writeHead(200,                              {"Content-Type":
"text/plain;charset=utf-8"});
        response.write("Hello World");
        response.end();
    }
    //把函数当作参数传递
    http.createServer(onRequest).listen(8888);

    console.log("Server has started.");
    console.log("请在浏览器中打开 http://127.0.0.1:8888");
}

exports.start = start;
```

这样我们就在 start()函数里添加了 handle 参数，并且把 handle 对象作为第一个参数传递给了 route()回调函数。

然后我们相应地在 route.js 文件中修改 route()函数：

```
function route(pathname) {
  console.log("About to route a request for " + pathname);
}
```

```
exports.route = route;
```

以上处理之后,就把服务器、路由和请求处理程序关联在一起了。现在我们启动应用程序,并在浏览器中访问 http://127.0.0.1:8888/start,以下日志可以说明系统调用了正确的请求处理程序:

```
node index.js
Server has started.
请在浏览器中打开 http://127.0.0.1:8888
Request for /start received.
About to route a request for /start
Request handler 'start' was called.
```

在浏览器中打开 http://127.0.0.1:8888/,可以看到这个请求同样被 start 请求处理程序处理了:

```
node index.js
Server has started.
请在浏览器中打开 http://127.0.0.1:8888
Request for / received.
About to route a request for /
Request handler 'start' was called.
```

10.5 请求处理程序

在服务端需要对接收到的请求做出响应,因此,需要让请求处理程序能够像 onRequest 函数那样可以和浏览器请求进行对话。

修改 requestHandler.js 文件内容如下:

```
function start() {
  console.log("Request handler 'start' was called.");
  return "Hello Start";
}

function upload() {
  console.log("Request handler 'upload' was called.");
  return "Hello Upload";
}

exports.start = start;
exports.upload = upload;
```

同时,请求路由需要将请求处理程序返回给它的信息返回给服务器。因此,我们需要对

router.js 做相应的修改：

```
function route(handle, pathname) {
  console.log("About to route a request for " + pathname);
  if (typeof handle[pathname] === 'function') {
    return handle[pathname]();
  } else {
    console.log("No request handler found for " + pathname);
    return "404 Not found";
  }
}

exports.route = route;
```

若请求有路由，则调用对应的处理函数；若请求无法路由，则返回相关的错误信息。

最后，需要对 server.js 进行重构，使得它能够将请求处理程序通过请求路由返回的内容响应给浏览器，代码如下：

```
let http = require("http");

let url = require("url");

//用一个函数将之前的内容包裹起来
let start = (route, handle) => {
        //箭头函数
    let onRequest = (request, response) => {

        let pathname = url.parse(request.url).pathname;
        console.log("Request for " + pathname + " received.");
        route(handle,pathname);

        response.writeHead(200, {"Content-Type":
"text/plain;charset=utf-8"});
        var content = route(handle, pathname)
        response.write(content);
        response.end();
    }
    //把函数当作参数传递
    http.createServer(onRequest).listen(8888);

    console.log("Server has started.");
}

exports.start = start;
```

运行重构后的应用，访问 http://localhost:8888/start，浏览器会输出 Hello Start；访问

http://localhost:8888/upload 会输出 Hello Upload；访问 http://localhost:8888/foo 会输出 404 Not found，如图 10.5 所示。

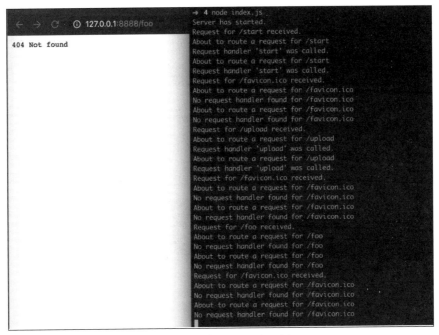

图 10.5 处理请求程序示例

但是，当有请求处理程序需要进行非阻塞的操作的时候，该应用就有一些问题。将 requestHandlers.js 修改成如下形式：

```
function start() {
  console.log("Request handler 'start' was called.");

  function sleep(milliSeconds) {
    var startTime = new Date().getTime();
    while (new Date().getTime() < startTime + milliSeconds);
  }

  sleep(10000);
  return "Hello Start";
}

function upload() {
  console.log("Request handler 'upload' was called.");
  return "Hello Upload";
}

exports.start = start;
```

```
exports.upload = upload;
```

上述代码中，先调用 upload()，会立即返回结果。当调用函数 start() 时，Node.js 会先等待 10 秒，之后才会返回"Hello Start"。图 10.6 所示为等待中。

图 10.6　阻塞式请求

分别在两个浏览器窗口中同时打开 http://localhost:8888/start 和 http://localhost:8888/upload 时，会发现 start 和 upload 都等待了 10 秒，原因就是 start() 包含阻塞操作，阻塞了所有其他的处理工作，如图 10.7 所示。

图 10.7　阻塞式请求

此问题需要使用非阻塞式处理请求的方式进行响应，详情请看下一节。

10.6　非阻塞式处理请求响应

Node.js 是单线程的，通过事件轮询（Event Loop）来实现并行操作，Node.js 可以在不新增额外线程的情况下，依然可以对任务进行并行处理。

到目前为止，通过前几节的介绍，示例应用已经可以通过应用各层之间传递值的方式，将请求处理程序返回的最终要显示给用户的内容传递给 HTTP 服务器。

之前介绍的是采用将内容传递给服务器的方式，现在介绍一种实现方式，采用将服务器"传递"给内容的方式，即将从服务器的回调函数 onRequest() 获取的 response 对象通过请求路由传递给请求处理程序，这之后处理程序就可以采用该对象上的函数来对请做作出响应。原理大致如此，接下来让我们来一步一步实现这种方案。

先从 server.js 开始编写代码：

```
const http = require("http");

const url = require("url");

//用一个函数将之前的内容包裹起来
let start = (route,handle) => {
    //箭头函数
```

```javascript
    let onRequest = (request, response) => {

        let pathname = url.parse(request.url).pathname;
        console.log("Request for " + pathname + " received.");
        route(handle, pathname, response);
    }
    //把函数当作参数传递
    http.createServer(onRequest).listen(8888);

    console.log("Server has started.");
}

exports.start = start;
```

相对此前从 route()函数获取返回值的做法，本示例将 response 对象作为第 3 个参数传递给 route()函数，并且将 onRequest()处理程序中所有有关 response 的函数都移除，因为本示例希望把这部分工作交给 route()函数来完成。

接下来对 router.js 进行修改：

```javascript
function route(handle, pathname, response) {
  console.log("About to route a request for " + pathname);
  if (typeof handle[pathname] === 'function') {
   handle[pathname](response);
  } else {
   console.log("No request handler found for " + pathname);
   response.writeHead(404, {"Content-Type": "text/plain"});
   response.write("404 Not found");
   response.end();
  }
}

exports.route = route;
```

同样的模式，相对此前从请求处理程序中获取返回值，这次取而代之的是直接传递 response 对象。

如果没有对应的请求处理器处理，我们就直接返回 404 错误。

最后，将 requestHandler.js 修改为如下形式：

```javascript
var exec = require("child_process").exec;

function start(response) {
  console.log("Request handler 'start' was called.");

  exec("ls -lah", function (error, stdout, stderr) {
    response.writeHead(200, {"Content-Type": "text/plain"});
```

```
    response.write(stdout);
    response.end();
  });
}

function upload(response) {
  console.log("Request handler 'upload' was called.");
  response.writeHead(200, {"Content-Type": "text/plain"});
  response.write("Hello Upload");
  response.end();
}

exports.start = start;
exports.upload = upload;
```

处理程序函数需要接收 response 参数,为了对请求做出直接的响应。start 处理程序在 exec() 的匿名回调函数中做请求响应的操作,而 upload 处理程序仍然是简单地回复 "Hello World",只是这次是使用 response 对象。

通过 node index.js 命令启动应用:

```
node index.js
Request for /start received.
About to route a request for /start
Request handler 'start' was called.
```

在浏览器中打开 http://127.0.0.1:8888/start,效果如图 10.8 所示。

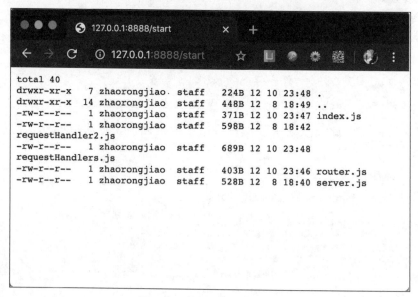

图 10.8 传递 response 对象

如果想要证明/start 处理程序中耗时的操作不会阻塞对/upload 请求做出立即响应的话,就可以将 requestHandlers.js 修改为如下形式：

```javascript
var exec = require("child_process").exec;

function start(response) {
  console.log("Request handler 'start' was called.");

  exec("find /",
    { timeout: 10000, maxBuffer: 20000*1024 },
    function (error, stdout, stderr) {
      response.writeHead(200, {"Content-Type": "text/plain"});
      response.write(stdout);
      response.end();
    });
}

function upload(response) {
  console.log("Request handler 'upload' was called.");
  response.writeHead(200, {"Content-Type": "text/plain"});
  response.write("Hello Upload");
  response.end();
}

exports.start = start;
exports.upload = upload;
```

当请求 http://localhost:8888/start 时，需要持续等待 10 秒钟之后页面才能加载完成，而当请求 http://localhost:8888/upload 时，即使这个时候/start 响应还在处理中，处理程序仍然会立即响应。

10.7 处理 POST 请求

本节介绍如何处理 POST 请求。使用 POST 请求的场景有很多，一个常见的场景是显示一个文本区（Textarea）供用户输入较长篇幅的内容，然后通过 POST 请求提交给服务器。服务器接收到请求之后，通过处理程序将输入的内容展示到浏览器中。

在本示例中，通过/start 请求处理程序生成供用户输入的带文本区的表单，因此，将 requestHandlers.js 修改为如下形式：

```javascript
// 引入 child_process 模块
var exec = require("child_process").exec;

function start(response) {
```

```
  console.log("Request handler 'start' was called.");

  let body = '<html>'+
    '<head>'+
    '<meta http-equiv="Content-Type" content="text/html; '+
    'charset=UTF-8" />'+
    '</head>'+
    '<body>'+
    '<form action="/upload" method="post">'+
    '<textarea name="text" rows="5" cols="60"></textarea>'+
    '<input type="submit" value="Submit text" />'+
    '</form>'+
    '</body>'+
    '</html>';

  response.writeHead(200, {"Content-Type": "text/html;charset=utf-8"});
  response.write(body);
  response.end();
}

function upload(response) {
  console.log("Request handler 'upload' was called.");
  response.writeHead(200, {"Content-Type": "text/plain;charset=utf-8"});
  response.write("Hello Upload");
  response.end();
}

exports.start = start;
exports.upload = upload;
```

引入了一个新的 Node.js 模块 child_process。之所以用它，是为了实现一个既简单又实用的非阻塞操作 exec()。在浏览器中请求 http://127.0.0.1:8888/start，效果如图 10.9 所示。

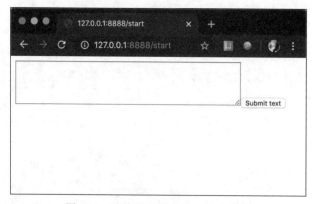

图 10.9　渲染用于用户输入的文本框

当用户提交表单时，将触发/upload 请求处理程序处理 POST 请求的问题。用户可能会输入大量的内容，用阻塞的方式处理大数据量的请求必然会导致用户操作的阻塞。为了使整个过程非阻塞，Node.js 会将 POST 数据拆分成很多小的数据块，然后通过触发特定的事件将这些小数据块传递给回调函数。这里的特定事件有 data 事件（表示新的小数据块到达了）以及 end 事件（表示所有的数据都已经接收完毕）。需要告诉 Node.js 当这些事件触发的时候回调哪些函数，这个过程是通过在 request 对象上注册监听器（Listener）来实现的。这里的 request 对象在每次接收到 HTTP 请求时，都会把该对象传递给 onRequest 回调函数。

示例代码如下：

```
request.addListener("data", function(chunk) {
  // called when a new chunk of data was received
});

request.addListener("end", function() {
  // called when all chunks of data have been received
});
```

将 data 和 end 事件的回调函数直接放在服务器中，在 data 事件回调中收集所有的 POST 数据，当接收到所有数据，触发 end 事件后，其回调函数调用请求路由，并将数据传递给它，然后请求路由再将该数据传递给请求处理程序。

建议此时直接在服务器中处理 POST 数据，然后将最终的数据传递给请求路由和请求处理器来做进一步的处理。将 server.js 修改如下：

```
const http = require("http");
const url = require("url");

const start = (route, handle) => {
  //箭头函数
  let onRequest = (request, response) => {

    let postData = "";
    let pathname = url.parse(request.url).pathname;
    console.log("Request for " + pathname + " received.");
    // 设置接收数据的编码格式
    request.setEncoding("utf8");
    // 注册 data 事件监听器
    request.addListener("data", function (postDataChunk) {
      // 将接收到的数据赋值给 postData 变量
      postData += postDataChunk;
      // 输出日志
      console.log("Received POST data chunk '" + postDataChunk + "'.");
    });
    // 注册 end 事件监听器
```

```
    request.addListener("end", function () {
      // 将 postData 传递给请求路由
      route(handle, pathname, response, postData);
    });
  }
  //把函数当作参数传递
  http.createServer(onRequest).listen(8888);

  console.log("Server has started.");
}

exports.start = start;
```

上述代码中，首先设置了接收数据的编码格式为 UTF-8；然后注册了 data 事件的监听器，用于收集每次接收到的新数据块，并将其赋值给 postData 变量；最后将请求路由的调用移到 end 事件处理程序中，以确保它只会当所有数据接收完毕后才触发，并且只触发一次，同时把 POST 数据传递给请求路由，请求处理程序会使用 POST 数据做进一步的展示处理。

上述代码在每个数据块到达的时候输出了日志，在开发阶段是很有用的，有助于开发者准确定位当前发生了什么。但是在最终生产环境中，有可能会在数据量很大的情况下发生异常，需要移除这部分日志代码。尝试着去输入一小段文本或大段内容，当输入大段内容的时候，就会发现 data 事件会触发多次。

接下来，在/upload 页面展示用户输入的内容。要实现该功能，需要将 postData 传递给请求处理程序，将 router.js 修改如下：

```
unction route(handle, pathname, response, postData) {
  console.log("About to route a request for " + pathname);
  if (typeof handle[pathname] === 'function') {
    handle[pathname](response, postData);
  } else {
    console.log("No request handler found for " + pathname);
    response.writeHead(404, {"Content-Type": "text/plain"});
    response.write("404 Not found");
    response.end();
  }
}

exports.route = route;
```

然后，在 requestHandlers.js 中将数据包含在对 upload 请求的响应中：

```
function start(response, postData) {
  console.log("Request handler 'start' was called.");

  var body = '<html>'+
```

```
      '<head>'+
      '<meta http-equiv="Content-Type" content="text/html; '+
      'charset=UTF-8" />'+
      '</head>'+
      '<body>'+
      '<form action="/upload" method="post">'+
      '<textarea name="text" rows="20" cols="60"></textarea>'+
      '<input type="submit" value="Submit text" />'+
      '</form>'+
      '</body>'+
      '</html>';

    response.writeHead(200, {"Content-Type": "text/html"});
    response.write(body);
    response.end();
}

function upload(response, postData) {
    console.log("Request handler 'upload' was called.");
    response.writeHead(200, {"Content-Type": "text/plain"});
    response.write("You've sent: " + postData);
    response.end();
}

exports.start = start;
exports.upload = upload;
```

修改之后就可以接收POST数据并在请求处理程序中处理接收到的数据了。当前示例是把请求的整个消息体传递给了请求路由和请求处理程序，在实际场景中，应该只把POST数据中感兴趣的部分传递给请求路由和请求处理程序。在本次示例中，真实有用的数据只有text字段，可以使用querystring模块来实现，示例代码如下：

```
const querystring = require("querystring");

function start(response, postData) {
    console.log("Request handler 'start' was called.");

    const body = '<html>' +
      '<head>' +
      '<meta http-equiv="Content-Type" content="text/html; ' +
      'charset=UTF-8" />' +
      '</head>' +
      '<body>' +
      '<form action="/upload" method="post">' +
      '<textarea name="text" rows="20" cols="60"></textarea>' +
```

```
   '<input type="submit" value="Submit text" />' +
   '</form>' +
   '</body>' +
   '</html>';

  response.writeHead(200, {
    "Content-Type": "text/html"
  });
  response.write(body);
  response.end();
}

function upload(response, postData) {
  console.log("Request handler 'upload' was called.");
  response.writeHead(200, {
    "Content-Type": "text/plain"
  });
  response.write("You've sent the text: " + querystring.parse(postData).text);
  response.end();
}

exports.start = start;
exports.upload = upload;
```

使用 node index.js 命令启动应用程序，在浏览器中访问 http://127.0.0.1:8888/start，效果如图 10.10 所示。

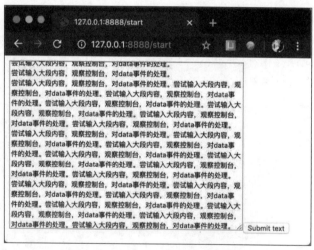

图 10.10　POST 请求的数据输入

单击 Submit text 按钮，将跳转到 http://127.0.0.1:8888/upload，效果如图 10.11 所示。

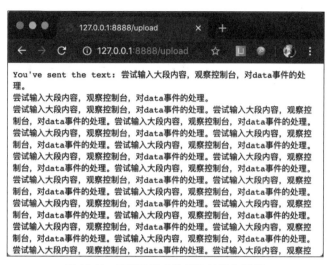

图 10.11　POST 请求数据的处理与展示

10.8 文件上传

本节介绍如何从本地上传图片文件，服务端接收之后，返回给客户端，并将该图片在浏览器中显示出来。同时，介绍如何安装外部 Node.js 模块并在应用中使用。

本示例需要使用的外部模块是 formidable 模块，该模块对解析上传的文件数据处理进行抽象。事实上，处理文件上传也是处理 POST 数据的一种。

首先安装需要使用的 formidable 模块，使用 Node.js 包管理器 NPM 进行安装即可。在安装模块之前，在项目路径使用 npm init 命令创建 package.json 文件，新增的 package.json 文件的内容如下：

```
{
  "author" : "",
  "description" : "",
  "license" : "ISC",
  "main" : "index.js",
  "name" : "10-8",
  "scripts" : {
    "test" : "echo \"Error: no test specified\" && exit 1"
  },
  "version" : "1.0.0"
}
```

接下来，在终端输入如下命令安装 formidable 外部模块：

```
npm install formidable --save
```

过程如下:

```
npm install formidable --save
npm notice created a lockfile as package-lock.json. You should commit this file.
npm WARN 10-8@1.0.0 No description
npm WARN 10-8@1.0.0 No repository field.

+ formidable@1.2.1
removed 1 package, updated 1 package and audited 1 package in 6.133s
found 0 vulnerabilities
```

之后 package.json 文件更新后如下:

```
{
  "author": "liyuechun",
  "description": "",
  "license": "ISC",
  "main": "index.js",
  "name": "fileupload",
  "scripts": {
    "test": "echo \"Error: no test specified\" && exit 1"
  },
  "version": "1.0.0",
  "dependencies": {
    "formidable": "^1.1.1"
  }
}
```

此时,formidable 模块安装完毕后,就可以在项目中使用 formidable 模块了。使用外部模块与内部模块类似,用 require 语句将其引入即可:

```
const formidable = require("formidable");
```

该模块的作用是使得通过 HTTP POST 请求提交的表单,在 Node.js 中可以被解析。项目中需要做的是创建一个新的 IncomingForm,IncomingForm 是对提交表单的抽象表示。创建完成之后,可以用它解析 request 对象,获取表单中需要的数据字段。

formidable 官方的例子展示了这两部分是如何融合在一起工作的:

```
let formidable = require('formidable'),
    http = require('http'),
    util = require('util');

http.createServer(function(req, res) {
  if (req.url == '/upload' && req.method.toLowerCase() == 'post') {
    // parse a file upload
    let form = new formidable.IncomingForm();
```

```
    form.parse(req, function(err, fields, files) {
      res.writeHead(200, {'content-type': 'text/plain'});
      res.write('received upload:\n\n');
      res.end(util.inspect({fields: fields, files: files}));
    });
    return;
  }

  // show a file upload form
  res.writeHead(200, {'content-type': 'text/html'});
  res.end(
    '<form action="/upload" enctype="multipart/form-data" '+
    'method="post">'+
    '<input type="text" name="title"><br>'+
    '<input type="file" name="upload" multiple="multiple"><br>'+
    '<input type="submit" value="Upload">'+
    '</form>'
  );
}).listen(8888);
```

将上述代码保存到一个文件中，并通过 Node 来执行，就可以进行简单的表单提交了（包括文件上传）。然后，可以看到通过调用 form.parse 传递给回调函数的 files 对象的内容，如下所示：

```
{ fields: { title: 'Hello World' },
  files:
   { upload:
      { size: 1558,
        path: './tmp/1c747974a27a6292743669e91f29350b',
        name: 'us-flag.png',
        type: 'image/png',
        lastModifiedDate: Tue, 21 Jun 2011 07:02:41 GMT,
        _writeStream: [Object],
        length: [Getter],
        filename: [Getter],
        mime: [Getter] } } }
```

为了实现目标功能，将上述代码应用到示例应用中。另外，还要考虑如何将上传文件的内容保存在./tmp 目录中，并显示在浏览器上。首先需要将该图片保存到 tmp 目录下，并使用 fs 模块将该文件读取到服务器。添加/showURL 请求处理程序，该处理程序直接硬编码将文件./tmp/test.png 内容展示到浏览器中。将 requestHandlers.js 修改为如下形式：

```
var querystring = require("querystring"),
    fs = require("fs");

function start(response, postData) {
```

```
  console.log("Request handler 'start' was called.");

  var body = '<html>'+
    '<head>'+
    '<meta http-equiv="Content-Type" '+
    'content="text/html; charset=UTF-8" />'+
    '</head>'+
    '<body>'+
    '<form action="/upload" method="post">'+
    '<textarea name="text" rows="20" cols="60"></textarea>'+
    '<input type="submit" value="Submit text" />'+
    '</form>'+
    '</body>'+
    '</html>';

  response.writeHead(200, {"Content-Type": "text/html"});
  response.write(body);
  response.end();
}

function upload(response, postData) {
  console.log("Request handler 'upload' was called.");
  response.writeHead(200, {"Content-Type": "text/plain"});
  response.write("You've sent the text: "+
  querystring.parse(postData).text);
  response.end();
}

function show(response, postData) {
  console.log("Request handler 'show' was called.");
  fs.readFile("./tmp/test.png", "binary", function(error, file) {
    if(error) {
      response.writeHead(500, {"Content-Type": "text/plain"});
      response.write(error + "\n");
      response.end();
    } else {
      response.writeHead(200, {"Content-Type": "image/png"});
      response.write(file, "binary");
      response.end();
    }
  });
}

exports.start = start;
exports.upload = upload;
```

```
exports.show = show;
```

还需要将新的请求处理程序添加到 index.js 的路由映射表中:

```
//从'server'模块中导入 server 对象

let server = require('./server');
let router = require("./router");
let requestHandlers = require("./requestHandlers");

//对象构造
var handle = {}
handle["/"] = requestHandlers.start;
handle["/start"] = requestHandlers.start;
handle["/upload"] = requestHandlers.upload;
handle["/show"] = requestHandlers.show;

//启动服务器
server.start(router.route, handle);
```

修改 server.js,将 request 对象传递给请求路由处理器:

```
const http = require("http");
const url = require("url");

//用一个函数将之前的内容包裹起来
let start = (route,handle) => {
        //箭头函数
    let onRequest = (request, response) => {

        let pathname = url.parse(request.url).pathname;
        console.log("Request for " + pathname + " received.");
        route(handle, pathname, response, request);
    }
    //把函数当作参数传递
    http.createServer(onRequest).listen(8888);

    console.log("Server has started.");
}

exports.start = start;
```

router.js 不再需要传递 postData,传递 request 对象即可:

```
function route(handle, pathname, response, request) {
  console.log("About to route a request for " + pathname);
  if (typeof handle[pathname] === 'function') {
    handle[pathname](response, request);
  } else {
    console.log("No request handler found for " + pathname);
    response.writeHead(404, {"Content-Type": "text/html"});
```

```
      response.write("404 Not found");
      response.end();
    }
}

exports.route = route;
```

现在，request 对象就可以在 upload 请求处理程序中使用了。formidable 会将上传的文件保存到本地/tmp 目录中。这里采用 fs.renameSync(path1,path2)来实现。要注意的是，该方法是同步执行的，也就是说如果该重命名的操作很耗时，就会发生阻塞。

接下来，把处理文件上传以及重命名的操作放到一起，修改后的 requestHandlers.js 如下：

```
var querystring = require("querystring"),
    fs = require("fs"),
    formidable = require("formidable");

function start(response) {
  console.log("Request handler 'start' was called.");

  var body = '<html>'+
    '<head>'+
    '<meta http-equiv="Content-Type" content="text/html; '+
    'charset=UTF-8" />'+
    '</head>'+
    '<body>'+
    '<form action="/upload" enctype="multipart/form-data" '+
    'method="post">'+
    '<input type="file" name="upload" multiple="multiple">'+
    '<input type="submit" value="Upload file" />'+
    '</form>'+
    '</body>'+
    '</html>';

    response.writeHead(200, {"Content-Type": "text/html"});
    response.write(body);
    response.end();
}

function upload(response, request) {
  console.log("Request handler 'upload' was called.");

  var form = new formidable.IncomingForm();
  console.log("about to parse");
  form.parse(request, function(error, fields, files) {
    console.log("parsing done");
    fs.renameSync(files.upload.path, "./tmp/test.png");
    response.writeHead(200, {"Content-Type": "text/html"});
    response.write("received image:<br/>");
    response.write("<img src='/show' />");
```

```
    response.end();
  });
}

function show(response) {
  console.log("Request handler 'show' was called.");
  fs.readFile("./tmp/test.png", "binary", function(error, file) {
    if(error) {
      response.writeHead(500, {"Content-Type": "text/plain"});
      response.write(error + "\n");
      response.end();
    } else {
      response.writeHead(200, {"Content-Type": "image/png"});
      response.write(file, "binary");
      response.end();
    }
  });
}

exports.start = start;
exports.upload = upload;
exports.show = show;
```

重启服务器，在浏览器中访问 http://127.0.0.1:8888，效果如图 10.12 和图 10.13 所示。

图 10.12　上传文件示例

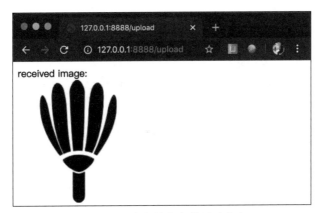

图 10.13　将上传的文件显示出来

第 11 章
实战：基于 Vue+Express+MongoDB实现一个后台管理系统

本章介绍一个后台管理项目，核心功能包括登录、增删改查涉及的前端和后端代码的各自实现以及前后端的交互。项目整体比较简单，但是是一次比较全的实践，涉及 Vue、Express、MongoDB 等框架技术，一次完整的前后端结合的项目实践还是比较有意义的。同时，注意以上示例代码在项目环境中打印的日志信息，在生产环境部署时，需要删除这些日志信息，避免对线上稳定性产生意料之外的影响。

对 Node.js 开发环境不熟悉的同学，可参考第 1 章关于如何在不同的操作系统中安装 Node.js 的介绍，准备 Node.js 的开发环境。

11.1 项目结构

项目的目录结构明细如下：

```
├── build                          // 存放构建文件
│   ├── build.js                   // 构建代码，执行 webpack 编译任务
│   ├── check-versions.js          // 版本
│   ├── utils.js                   // 工具
│   ├── webpack.base.conf.js       // webpack 基本通用配置
│   ├── webpack.dev.conf.js        // webpack 开发环境配置
│   └── webpack.prod.conf.js       // webpack 生产环境配置
├── config                         // 项目参数配置的文件
│   ├── db.js                      // 数据库配置
│   ├── dev.env.js                 // 环境配置
│   ├── prod.env.js                // 路由页面
│   └── index.js                   // 基础配置
├── models
```

```
|       ├── heroSchema.js          // hero 数据结构
|       └── userSchema.js          // user 数据结构
├── node_modules                   // 项目依赖模块
├── router
|       ├── hero.js                // hero 路由管理
|       └── user.js                // user 路由管理
├── src
|       ├── assets                 // 静态资源文件
|       ├── components             // 前端基础组件
|       ├── ruter                  // 路由页面
|       ├── utils                  // 工具函数
|       ├── App.vue                // 应用
|       └── main.js                // 启动前端应用
├── static                         // 项目静态文件的入口
├── .babelrc
├── .editorconfig
├── .gitignore
├── .postcssrc.js
├── app.js                         // 启动后端应用
├── index.html
├── package.json
└── README.md
```

其中，build 文件夹存放 Webpack 的配置文件，src 中为前端代码，config 中存放项目参数配置的文件，static 为项目静态文件的入口，node_modules 存放项目依赖模块，.babelrc 是 babel 编译配置，.postcssrc 是 CSS 后处理器的配置。

11.2 前端代码实现

11.2.1 项目依赖的模块

本章项目依赖的模块：

- vue/vue-router/vuex：Vue 全家桶。
- axios：基于 promise 的 HTTP 库，可以用在浏览器和 Node.js 中。
- qs：用于解决 axios POST 请求参数的问题。
- element-ui：VUE 2.0 UI 框架。
- babel-polyfill：用于实现浏览器不支持原生功能的代码。
- highlight.js/marked：两者搭配实现 Markdown 的常用语法。

11.2.2 注册页

一般后台管理系统都要求登录之后才允许查看内容或者进行相关操作,注册是后台管理系统的基本功能之一。

注册页的核心功能是提供一个表单,表单中包含便于用户输入的账号、密码、密码确认等基础信息。

```
<template>
  <div class="register-page">
    <div class="form">
      <p class="form-title">用户注册</p>
      <div class="form-item">
        <span class="form-item-title">账号</span>
        <input class="form-input" type="text" placeholder="请输入账号" v-model="userInfo.username" />
      </div>
      <div class="form-item">
        <span class="form-item-title">密码</span>
        <input class="form-input" type="password" placeholder="请输入密码" v-model="userInfo.password" />
      </div>
      <div class="form-item">
        <span class="form-item-title">确认密码</span>
        <input
          class="form-input"
          type="password"
          placeholder="请再次输入密码"
          v-model="userInfo.re_password"
        />
      </div>
      <div class="form-item">
        <button class="form-submit-btn" @click="submit">注册</button>
      </div>
      <div class="form-bottom">
        <router-link class="form-jump-link" to="/login">已有账号,去登录</router-link>
      </div>
    </div>
  </div>
</template>

<script>
import Vue from "vue";
import request from "@/utils/request";

export default {
  data() {
    return {
```

```js
      userInfo: {
        username: "",
        password: "",
        re_password: ""
      }
    };
  },
  methods: {
    submit() {
      const {
        username,
        password,
        re_password
      } = this.userInfo
      if (
        username == "" || password == ""
      ) {
        Vue.$toast.error("注册失败,请填写完整表单", {
          position: "top"
        });
        return;
      }
      if (password.length < 5) {
        Vue.$toast.error("注册失败,密码最少为5位", {
          position: "top"
        });
        return;
      }
      if (password != re_password) {
        Vue.$toast.error("注册失败,2次密码输入不一致!", {
          position: "top"
        });
        return;
      }
      request({
        url: './register',
        method: "post",
        data: {
          username,
          password
        }
      })
        .then(response => {
          console.log('response', response)
          this.$router.push({path:"/list"})
        })
        .catch(error => {
          console.log(error);
        });
```

```
      }
    }
  };
</script>

<style lang="scss" rel="stylesheet/scss">
.register-page {
  padding-top: 40px;
}
.form {
  width: 600px;
  margin: auto;
  box-sizing: border-box;
  box-shadow: 0 0 30px #ccc;
  padding: 0 30px 0 15px;
  .form-title {
    font-size: 26px;
    font-weight: bold;
    height: 80px;
    line-height: 80px;
  }
  .form-item {
    height: 40px;
    padding: 10px 0;
    display: flex;
    justify-content: center;
  }
  .form-item-title {
    width: 120px;
    display: block;
    line-height: 40px;
    text-align: right;
    box-sizing: border-box;
    padding-right: 20px;
  }
  .form-input {
    flex: 1;
    font-size: 16px;
    box-sizing: border-box;
    padding: 0 5px;
    border: none;
    border-bottom: 1px dotted #ccc;
    outline: none;
    transition: border-bottom-color 0.3s;
    &:focus {
      border-bottom-color: red;
    }
  }
  .form-checkcode {
```

```
    width: 120px;
    cursor: pointer;
  }
  .form-submit-btn {
    padding: 0 40px;
    font-size: 22px;
    cursor: pointer;
  }
  .form-bottom {
    height: 40px;
    padding: 10px 0;
    .form-jump-link {
      color: #666;
      text-decoration: none;
      transition: color 0.3s;
      &:hover {
        color: red;
      }
    }
  }
}
</style>
```

样式代码使用 scss 语法。运行 npm start 命令启动应用，在浏览器中打开 http://localhost:8081/#/register，效果如图 11.1 所示。

图 11.1 注册页

对于已有账号的用户，提供跳转到登录页的链接，通过 router-link 组件实现跳转到登录页的逻辑：

```
<div class="form-bottom">
  <router-link class="form-jump-link" to="/login">已有账号，去登录</router-link>
</div>
```

11.2.3 登录页

在未登录的情况下，访问任何页面都将被重定向到登录页面。而对于已经登录的用户，可以直接访问列表页。登录页模板代码如下：

```
<template>
  <div class="login-container">
    <el-form ref="form" :model="form" label-width="80px" :rules="loginRules" class="login-form">
      <h4 class="title">{{title}}</h4>

      <el-form-item prop="username">
        <el-input v-model="form.username" prefix-icon="iconfont myIcon-user" placeholder="账号"></el-input>
      </el-form-item>
      <el-form-item prop="password">
        <el-input
          v-model="form.password"
          prefix-icon="iconfont myIcon-password"
          placeholder="密码"
          :type="passwordType"
        >
          <i slot="suffix" class="iconfont myIcon-eye" @click="showPwd"></i>
        </el-input>
      </el-form-item>

      <div class="login">
        <el-button type="primary" @click="login" class="login-btn" :loading="isloading">登录</el-button>
      </div>
    </el-form>
  </div>
</template>
```

登录时，需要对用户的输入进行检测，包括是否为空、账号格式是否正确、调用后台登录逻辑等，示例代码如下（其中，login 接口将在下一小节后台功能部分进行详细介绍）：

```
<script>
import request from "@/utils/request";
import { isvalidUsername } from "@/utils/validate";
export default {
  name: "login-page",
  data: function() {
    // 判断账号是否合法
    const validateUsername = (rule, value, callback) => {
      if (!isvalidUsername(value)) {
```

```javascript
          callback(new Error("请输入正确的账号"));
        } else {
          callback();
        }
      };
      //判断密码是否合法
      const validatePass = (rule, value, callback) => {
        if (value.length < 5) {
          callback(new Error("密码不能小于5位"));
        } else {
          callback();
        }
      };
      return {
        title: "管理后台",
        loginUrl: "./login",
        form: {
          username: "",
          password: ""
        },
        loginRules: {
          username: [
            { required: true, trigger: "blur", validator: validateUsername }
          ],
          password: [{ required: true, trigger: "blur", validator: validatePass }]
        },
        passwordType: "password",
        isloading: false
      };
    },
    methods: {
      // 登录
      login() {
        var $this = this;
        this.$refs.form.validate(valid => {
          if (valid) {
            this.isloading = true;
            request({
              url: this.loginUrl,
              method: "post",
              data: this.form
            })
              .then(response => {
                //模拟异步请求时间
                setTimeout(function() {
```

```
          $this.isloading = false;
          $this.$router.push({ path: "/list" });
        }, 3000);
      })
        .catch(error => {
          console.log(error);
        });
    } else {
      console.log("不请求");
    }
  });
},
// 切换密码框
showPwd() {
  if (this.passwordType === "password") {
    this.passwordType = "";
  } else {
    this.passwordType = "password";
  }
}
},
mounted: function() {}
};
</script>
```

以下是 CSS 样式代码示例：

```
<style lang="scss" rel="stylesheet/scss">
.login-container {
  width: 100%;
  height: 100%;
  background-color: rgb(0, 29, 41);
  background-repeat: no-repeat;
  background-size: cover;
  position: fixed;
  color: #fff;
  .title {
    font-size: 20px;
    margin: 0;
    padding: 40px 0;
  }
  .el-input__inner {
    padding-left: 40px;
  }
```

```css
.el-form-item {
  margin-bottom: 40px;
}
.el-input {
  width: 84%;
  input {
    background: transparent;
    color: #fff;
    height: 50px;
  }
}
.login-form {
  position: absolute;
  left: 0;
  right: 0;
  top: 0;
  bottom: 0;
  width: 620px;
  padding: 35px 50px 15px;
  margin: 120px auto;
}
.title,
.login {
  text-align: center;
}
.login-btn {
  width: 150px;
}
.userinfo {
  text-align: center;
  font-style: italic;
  span {
    padding: 0 10px;
  }
}
}
</style>
```

登录页的效果如图 11.2 所示。

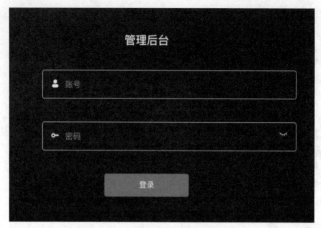

图 11.2　登录页展示效果示意图

11.2.4　管理页

管理页中包含管理后台的核心功能，包括基础的增加、删除、修改、搜索等操作，以及关键信息的展示。

1. 列表页

列表页提供查询操作及查询的结果，提供增删改查的操作按钮入口，并支持分页等功能，示例代码如下：

```html
<header>
  <label for="" class="formLabelCss">名称:</label>
  <el-input v-model="heroName" class="formInputCss" clearable placeholder="请输入英雄名称"></el-input>

  <label for="" class="formLabelCss">位置:</label>
  <el-select v-model="heroPosition" class="formInputCss">
    <el-option
      v-for="item in heroPositions"
      :key="item.value"
      :label="item.label"
      :value="item.value">
    </el-option>
  </el-select>

  <label for="" class="formLabelCss">性别:</label>
  <el-select v-model="heroSex" class="formInputCss">
    <el-option
      v-for="item in heroSexs"
      :key="item.value"
      :label="item.label"
```

```html
          :value="item.value">
      </el-option>
    </el-select>

    <el-button    type="primary"    class="searchBtn"    icon="el-icon-search"
@keyup.enter.native="search"  @click="search">查询</el-button>
    <el-button type="text" @click="clear">清空</el-button>

  </header>

  <nav>
    <el-button type="primary" class="addBtn" @click="add" icon="el-icon-plus">
添加</el-button>
  </nav>

  <main>
    <el-table
    :data="tableData"
    stripe
    element-loading-text="拼命加载中"
    header-row-class-name="tableHeader"
    v-loading.fullscreen.lock="loading"
    empty-text="亲，暂时没有数据哦"
    border
    style="width: 100%">
      <el-table-column
      fixed
      prop="heroName"
      label="英雄"
      align="center"
      width="120">
      </el-table-column>
      <el-table-column
      prop="age"
      label="年龄"
      align="center"
      width="100">
      </el-table-column>
        <el-table-column
        label="性别"
        align="center"
        width="100">
          <template slot-scope="scope">
            <i class="iconfont myIcon-man" v-if="scope.row.heroSex=='man'"></i>
            <i            class="iconfont            myIcon-woman"
```

```
v-if="scope.row.heroSex=='woman'"></i>
              {{jungleSex(scope.row.heroSex)}}
          </template>
      </el-table-column>
      <el-table-column
        prop="address"
        label="描述"
        align="center"
        width="150">
      </el-table-column>
      <el-table-column
        prop="dowhat"
        label="位置"
        align="center"
        width="163">
        <template slot-scope="scope">{{junglePosition(scope.row.heroPosition)}}</template>
      </el-table-column>
      <el-table-column
        prop="favourite"
        label="技能"
        align="center"
        width="200">
      </el-table-column>
      <el-table-column
        label="操作"
        align="center"
        width="460">
          <template slot-scope="scope">
            <el-button size="small" type="primary" @click="toDetail(scope.row['_id'])">详情</el-button>
            <el-button size="small" type="success" @click="modify(scope.row)">修改</el-button>
            <el-button type="danger" size="small" @click="deleteHero(scope.row['_id'])">删除</el-button>
            <el-button type="warning" size="small" @click="addPic(scope.row['_id'])">添加图片</el-button>
          </template>
      </el-table-column>
    </el-table>

    <el-pagination
      v-if="paginationShow"
      class="pagination"
      @current-change="handleCurrentChange"
```

```
    :current-page.sync="currentPage"
    :page-size="3"
    layout="total, prev, pager, next, jumper"
    :total="length">
  </el-pagination>
</main>
```

展示效果如图 11.3 所示。

图 11.3 列表页

2. 添加页

添加操作以弹窗形式展示，示例代码如下：

```
<!-- 新增数据 -->
  <el-dialog title="新增数据" :visible.sync="addFormVisible" class="addArea"
modal custom-class="addFormArea" @close="closeAdd('addForm')">
  <el-form  :model="addForm"  class="addForm"  :rules="rules"  status-icon
ref="addForm">

    <el-form-item label="名称:" :label-width="formLabelWidth" prop="heroName">
      <el-input v-model="addForm.heroName" auto-complete="off" placeholder="
请输入英雄名称"></el-input>
    </el-form-item>

    <el-form-item label="年龄:" :label-width="formLabelWidth" prop="age">
      <el-input v-model.number="addForm.age" auto-complete="off" placeholder="
请输入英雄年龄"></el-input>
    </el-form-item>

    <el-form-item label="性别:" :label-width="formLabelWidth" prop="heroSex">
      <el-select v-model="addForm.heroSex" class="sexArea" placeholder="请输入
英雄性别">
```

```html
            <el-option label="汉子" value="man"></el-option>
            <el-option label="妹子" value="woman"></el-option>
          </el-select>
        </el-form-item>

        <el-form-item label="描述:" :label-width="formLabelWidth" prop="address">
          <el-input v-model="addForm.address" auto-complete="off" placeholder="请输入英雄描述"></el-input>
        </el-form-item>

        <el-form-item label="位置:" :label-width="formLabelWidth" prop="heroPosition">
          <el-select v-model="addForm.heroPosition" placeholder="请选择英雄位置" class="sexArea" multiple>
            <el-option label="上单" value="miss"></el-option>
            <el-option label="打野" value="jungle"></el-option>
            <el-option label="中单" value="mid"></el-option>
            <el-option label="ADC" value="ad"></el-option>
            <el-option label="辅助" value="support"></el-option>
          </el-select>
        </el-form-item>

        <el-form-item label="技能:" :label-width="formLabelWidth" prop="favourite">
          <el-input v-model="addForm.favourite" auto-complete="off" placeholder="请输入英雄技能"></el-input>
        </el-form-item>

        <el-form-item label="技能介绍:" :label-width="formLabelWidth" prop="explain">
          <el-input v-model="addForm.explain" auto-complete="off" type="textarea" :rows="6" resize="none" placeholder="请输入英雄技能介绍" style="width: 520px"></el-input>
        </el-form-item>
      </el-form>
      <div slot="footer" class="dialog-footer">
        <el-button @click="addFormVisible = false">取 消</el-button>
        <el-button type="primary" @click="addSure('addForm')">确 定</el-button>
      </div>
    </el-dialog>
```

展示效果如图 11.4 所示。

图 11.4 添加页

3. 修改页

修改弹窗页与添加页类型以弹窗形式交互：

```
<!-- 修改数据 -->
<el-dialog title="修改数据" :visible.sync="modifyFormVisible" class="addArea"
modal custom-class="addFormArea" @close="closeModify('modifyForm')">
    <el-form :model="modifyForm" class="addForm" :rules="rules" status-icon
ref="modifyForm">

        <el-form-item label="英雄:" :label-width="formLabelWidth" prop="heroName">
        <el-input v-model="modifyForm.heroName" auto-complete="off"
placeholder="请输入英雄名称"></el-input>
        </el-form-item>

        <el-form-item label="年龄:" :label-width="formLabelWidth" prop="age">
        <el-input v-model.number="modifyForm.age" auto-complete="off"
placeholder="请输入英雄年龄"></el-input>
        </el-form-item>

        <el-form-item label="性别:" :label-width="formLabelWidth" prop="heroSex">
        <el-select v-model="modifyForm.heroSex" placeholder="请选择英雄性别"
class="sexArea">
            <el-option label="汉子" value="man"></el-option>
            <el-option label="妹子" value="woman"></el-option>
        </el-select>
        </el-form-item>
```

```html
        <el-form-item label="描述:" :label-width="formLabelWidth" prop="address">
            <el-input v-model="modifyForm.address" auto-complete="off" placeholder="请输入英雄描述"></el-input>
        </el-form-item>

        <el-form-item label="位置:" :label-width="formLabelWidth" prop="heroPosition">
            <el-select v-model="modifyForm.heroPosition" placeholder="请选择英雄位置" class="sexArea" multiple>
                <el-option label="上单" value="miss"></el-option>
                <el-option label="打野" value="jungle"></el-option>
                <el-option label="中单" value="mid"></el-option>
                <el-option label="ADC" value="ad"></el-option>
                <el-option label="辅助" value="support"></el-option>
            </el-select>
        </el-form-item>

        <el-form-item label="技能:" :label-width="formLabelWidth" prop="favourite">
            <el-input v-model="modifyForm.favourite" auto-complete="off" placeholder="请输入英雄技能"></el-input>
        </el-form-item>

        <el-form-item label="技能介绍:" :label-width="formLabelWidth" prop="explain">
            <el-input v-model="modifyForm.explain" auto-complete="off" type="textarea" :rows="6" resize="none" placeholder="请输入英雄背景"></el-input>
        </el-form-item>
    </el-form>
    <div slot="footer" class="dialog-footer">
        <el-button @click="modifyFormVisible = false">取 消</el-button>
        <el-button type="primary" @click="modifySure('modifyForm')">确 定</el-button>
    </div>
</el-dialog>

<!-- 添加图片 -->
<el-dialog title="添加图片" :visible.sync="addpicVisible" class="addPicArea" @close="closePicAdd">
    <el-form :model="addpicform">
        <el-form-item label="图片地址" :label-width="formLabelWidth">
            <el-input v-model="addpicform.url" auto-complete="off"></el-input>
        </el-form-item>
    </el-form>
    <div slot="footer" class="dialog-footer">
```

```
    <el-button @click="addpicVisible = false">取 消</el-button>
    <el-button type="primary" @click="addpicSure">确 定</el-button>
  </div>
</el-dialog>
```

示例效果如图 11.5 所示。

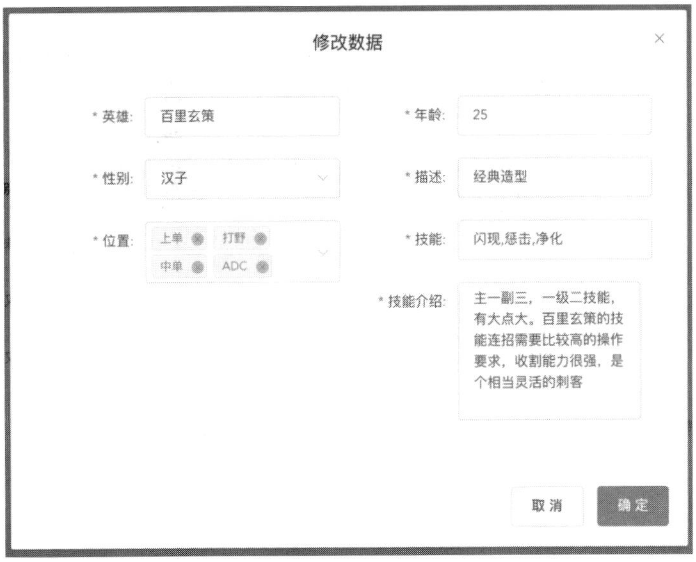

图 11.5 修改页

4. 详情页

详情页通过调用后台的查询接口获取单项的明细数据，最终使用相应的交互呈现给用户，模板代码示例如下：

```
<template>
  <div class="detail">
    <el-button type="success" class="goback" icon="el-icon-arrow-left" @click="goback">返回上一页</el-button>
    <el-carousel :interval="2000" type="card" height="300px" indicator-position="outside" v-if="imgArr.length">
      <el-carousel-item v-for="(item, index) in imgArr" :key="index">
        <img :src="item" alt class="imgClass" />
      </el-carousel-item>
    </el-carousel>
    <main class="clearfix">
      <section class="main-left">
        <h3 class="detail-title">召唤师名称：</h3>
        <p class="detail-introduct">{{name}}</p>
        <h3 class="detail-title">背景介绍：</h3>
        <p class="detail-introduct">{{explain}}</p>
      </section>
```

```
      </main>
    </div>
</template>
```

展示效果如图 11.6 所示。

图 11.6　详情页

11.3　后端代码实现

11.3.1　数据库设计

首先使用命令 use hero 创建 MongoDB 数据库：

```
> mongo
> use hero
switched to db hero
> show dbs
admin   0.000GB
config  0.000GB
hero    0.000GB
local   0.000GB
test    0.000GB
```

本示例核心涉及两个表，即用户表、英雄表，对应的表结构设计如下：

```
// 用户
const userSchema = new Schema({
  user_name: String,
  user_id: String,
  user_pwd: String,
```

```
    avatar: {
      type: String,
      default: ""
    }
})

// 英雄
const heroSchema = new Schema({
  heroName :String,
  age : Number,
  heroSex : String,
  address : String,
  heroPosition : [],
  imgArr:[],
  favourite:String,
  explain:String,
})
```

类似地，使用 mongoose 进行实现，userSchema.js 示例代码如下：

```
const mongoose = require('mongoose')

const userSchema = mongoose.Schema({
  username: String,
  password: String,
}, { collection: 'user'})

const User = module.exports = mongoose.model('User',userSchema);
```

heroSchema.js 代码示例如下：

```
const mongoose = require('mongoose')

const heroSchema = mongoose.Schema({
  heroName :String,
  age : Number,
  heroSex : String,
  address : String,
  heroPosition : [],
  imgArr:[],
  favourite:String,
  explain:String,
}, { collection: 'myhero'})

const Hero = module.exports = mongoose.model('hero', heroSchema);
```

11.3.2 启动应用

本示例介绍使用 Express 创建应用，通过 mongoose 操作数据库，注册 hero 和 user 两个接口路由，后台接口运行在端口 3000 上：

```javascript
const express = require('express');
const hero = require('./router/hero');
const user = require('./router/user');
const mongoose = require("mongoose");
const bodyParser = require("body-parser");
const cookieParser = require('cookie-parser');
//连接数据库，hero 为数据库名称，不是表名
const db = mongoose.connect('mongodb://localhost:27017/hero');

const app = express()
app.use(bodyParser.json());
app.use(bodyParser.urlencoded({
  extended: false
}));
app.use(cookieParser());
app.use('/api', user)
app.use('/api', hero)
app.listen(3000, () => {
  console.log('app listening on port 3000.')
})
```

11.3.3 注册/登录接口

注册接口首先接收前端调用的 post 接口，获得传入的 username 和 password 参数，查询数据库中是否有重复用户，检测通过之后，通过调用 User.create 往 user 表中增加一条记录。在 router 文件夹下创建 user.js 文件，输入如下示例代码：

```javascript
//引入 express 模块
const express = require("express");
//定义路由级中间件
const router = express.Router();
//引入数据模型模块
const User = require("../models/userSchema");

// 用户注册
router.post("/register", (req, res) => {
  // 使用 User model 上的 create 方法储存数据
  console.log(req)
  const { username = '', password = ''} = req.body;
  User.create(req.body, (err, user) => {
```

```js
      if (username == '' || password == "") {
        res.json({
          code: 401,
          status: "fail",
          message: "注册失败，请填写完整表单!"
        })
      } else {
        // 判断 user 是否重复
        User.find({username, password})
        .then(result => {
          if (result.length != 0) {
            res.json({
              code: 409,
              status: "fail",
              message: '注册失败，登录账号重复了，换一个吧！'
            })
          } else if (err) {
            res.json({
              status: "fail",
              error: err
            });
          } else {
            res.json({
              status: "success",
              message: "新增成功"
            });
          }
        })
        .catch(err => {
          res.json({
            status: "fail",
            message: err
          });
        });
      }
  });

  console.log(req.body)
});
```

运行 npm start 和 node app.js 命令，打开 http://localhost:8081/#/register 页面，输入用户名、密码和确认密码并提交之后，在控制台中查看接口返回结果，如图 11.7 所示。

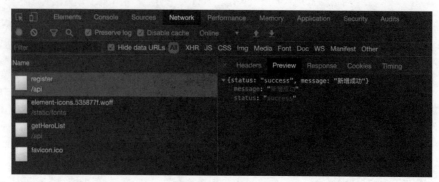

图 11.7　注册接口

类似地，登录接口首先获取 POST 请求体中的用户名和密码，并判断不为空，从数据库中查询对应的用户是否存在，若存在则返回登录成功的信息。在 router 文件夹下的 user.js 文件中补充如下代码：

```
//引入 express 模块
const express = require("express");
//定义路由级中间件
const router = express.Router();
//引入数据模型模块
const User = require("../models/userSchema");

//用户登录
router.post("/login", (req, res) => {
  console.log(req)
  let {username = '', password = ''} = req.body;
  if (username == '' || password == '') {
    res.json({
      code: 401,
      msg: "登录失败，请输入登录账号或密码！"
    });
  } else {
    User.find({username, password})
    .then(result => {
      if(result.length == 0){
        res.json({
          code: 401,
          msg: '登录失败，用户名或者密码错误！'
        });
      } else {
        res.json({
          code: 200,
          msg: "登录成功！",
          data: {
            _id: result[0]._id,
```

```
          user_name: result[0].username,
          avatar: result[0].avatar,
        }
      });
    }
  })
  .catch(err => {
    res.json({
      code: 200,
      msg: "登录成功!",
    });
  });
}
console.log(req.body)
});


module.exports = router;
```

运行 npm start 和 node app.js 命令,打开 http://localhost:8081/#/register 页面,输入用户名、密码和确认密码并提交之后,在控制台中查看接口返回结果,如图 11.8 所示。

图 11.8 登录接口

11.3.4 增删改查接口

首先介绍如何查询英雄信息列表。在 router 文件夹下创建 hero.js 文件,并注册 getHeroList 接口,针对传入的查询条件拼接查询 SQL,最后调用 find 进行查询,详细代码如下:

```
//引入 express 模块
const express = require("express");
//定义路由级中间件
const router = express.Router();
//引入数据模型模块
const Hero = require("../models/heroSchema");

// 查询所有英雄信息路由
```

```javascript
router.post("/getHeroList", (req, res) => {
  var heroPosition = new RegExp(req.body.heroPosition),
    heroSex = req.body.heroSex,
    heroName = req.body.heroName,
    pageNumber = req.body.pageNumber,
    pageRow = req.body.pageRow;

  // 根据查询入参个数动态生成sql查询语句
  var sqlObj = {};

  if (heroPosition) {
    sqlObj.heroPosition = heroPosition;
  }
  if (heroSex) {
    sqlObj.heroSex = heroSex;
  }
  if (heroName) {
    sqlObj.heroName = heroName;
  }
  var heroList = Hero.find(sqlObj);

  //对查询的结果进行筛选，skip跳过结果集中的前多少
  heroList.skip((pageNumber - 1) * pageRow);
  //对剩下来的数据限制返回个数
  heroList.limit(pageRow)

  // 实现分页的关键步骤
  heroList.exec(function (err, result) {
    if (err) {
      res.json({
        status: "fail",
        error: err
      });
    } else {
      Hero.find(sqlObj, function (err, heros) {
        res.json({
          status: "success",
          heroList: result,
          total: heros.length
        });
      })
    }
  })
});
```

在11.2.3小节中已经介绍了前端模板代码的编写，此时在前台代码中增加查询代码并将查

询结果显示出来：

```js
import request from "@/utils/request";

//查询
search() {
  this.paginationShow = false;

  var searchParmas = {
    heroName: this.heroName,
    heroPosition: this.heroPosition,
    heroSex: this.heroSex,
    pageNumber: this.pageNumber,
    pageRow: this.pageRow
  };
  this.loading = true;
  request({
    url: this.searchUrl,
    method: "post",
    data: searchParmas
  })
    .then(response => {
      this.$nextTick(function() {
        this.paginationShow = true;
      });
      this.loading = false;
      if (response.data.status == "success") {
        this.tableData = response.data.heroList;
        this.length = response.data.total;
      } else {
        this.tableData = [];
        this.$message({
          message: "查询出错，请重试!",
          type: "error"
        });
      }
    })
    .catch(error => {
      this.$nextTick(function() {
        this.paginationShow = true;
      });
      console.log(error);
    });

  //每次查询删除本地缓存
  sessionStorage.removeItem("queryParmas");
```

```
    }
```

其中，request 为自定义函数，但实际是使用 axios 封装的实现：

```
import axios from 'axios'
// 创建 axios 实例
const service = axios.create({
    baseURL: "http://localhost:8081/api", // api 的 base_url,相当于
http://localhost:3000
    timeout: 5000 // 请求超时时间
})

export default service
```

调用查询接口，在控制台查看接口返回的结果，如图 11.9 所示。

图 11.9　查询接口

再看增加接口的设计。在 router/hero.js 中注册 addHero 路由，获取 form 提交上来的字段数据，检测信息的完整性之后，调用 Hero.create()方法向数据库中存储数据：

```
// 添加一个英雄信息路由
router.post("/addHero", (req, res) => {
  console.log('addHero', req)
  //
{"heroName":"","age":"","heroSex":"","address":"","heroPosition":[],"favourite":"","explain":""}
  const { heroName = '', age = '', heroSex = '', address = '', heroPosition, favourite, explain} = req.body;
  // if (!heroName && !age && !heroSex && !address && !heroPosition && !favourite && !explain) {
    if (!heroName || !age) {
    res.json({
      status: "fail",
      message: '请填写完整英雄信息'
    });
    return
  }
```

```
// 使用 Hero model 上的 create 方法储存数据
Hero.create(req.body, (err, hero) => {
  if (err) {
    res.json({
      status: "fail",
      error: err
    });
  } else {
    res.json({
      status: "success",
      message: "新增成功!"
    });
  }
});

console.log(req.body)
});
```

在表单中输入信息，如图 11.10 所示。

图 11.10　新增英雄界面

单击"确定"按钮，在控制台查看 addHero 接口返回的结果，显示新增成功，如图 11.11 所示。

图 11.11 新增接口

在第 2 页的查询接口中可以看到刚才添加成功的数据,如图 11.12 所示。

图 11.12 查询新增的数据

修改数据库中已有的数据,在 router/hero.js 中增加 modifyHero 路由:

```javascript
//更新一条英雄信息数据路由
router.put("/modifyHero/:id", (req, res) => {
  console.log(req.params)
  Hero.findOneAndUpdate({
    _id: req.params.id
  }, {
    $set: {
      heroName: req.body.heroName,
      age: req.body.age,
      heroSex: req.body.heroSex,
      address: req.body.address,
      heroPosition: req.body.heroPosition,
      favourite: req.body.favourite,
      explain: req.body.explain
    }
  }, {
    new: true
  })
  .then(hero => res.json({
    status: "success",
    message: "修改成功"
  }))
  .catch(err => res.json({
```

```
      status: "fail",
      error: "修改失败"
    }));
});
```

在控制台查看接口,如图 11.13 所示。

图 11.13　修改接口

最后设计删除一条数据的接口,在 router/hero.js 中增加 deleteHero 路由:

```
//删除一条英雄信息路由
router.delete("/deleteHero/:id", (req, res) => {
  Hero.findOneAndRemove({
      _id: req.params.id
    })
    .then(hero => res.json({
      status: "success",
      message: "删除成功"
    }))
    .catch(err => res.json({
      status: "fail",
      message: "删除失败"
    }));
});
```

结果如图 11.14 所示,提示确定删除。

图 11.14　确定删除

删除接口返回结果如图 11.15 所示。

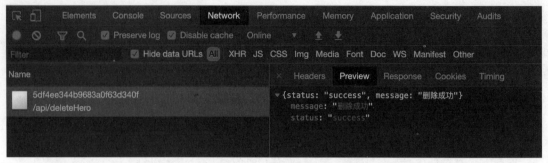

图 11.15　删除接口示意图

第 12 章

实战：基于Koa+MongoDB 实现博客网站

本章介绍如何使用 Node.js 实现完整的博客网站，包括所需的数据库组织结构、服务端接口、博客后台管理系统、博客的前台界面等相关内容。本章示例的服务端应用基于 Koa 实现。

提 示
Koa 是一个 Web 框架，可以方便地开发应用程序。

12.1 项目结构

为了方便组织调试，我们将博客的服务端、管理后台和前端站点的相关代码实现都在一个项目，本次实战示例的项目目录结构明细如下：

```
├── build                              // 存放构建文件
│   ├── build.js                       // 构建代码，执行 webpack 编译任务
│   ├── get-less-variables.js          // 管理 less 版本
│   ├── style-loader.js                // 样式构建配置项
│   ├── vue-config.js                  // 构建 Vue 配置项
│   ├── webpack.base.conf.js           // webpack 基本通用配置
│   ├── webpack.dev.conf.js            // webpack 开发环境配置
│   └── webpack.prod.conf.js           // webpack 生产环境配置
├── code
│   ├── admin                          // 管理后台
│   │   ├── src
│   │   │   ├── components             // 前端基础组件
│   │   │   ├── filters                // 工具函数
│   │   │   ├── images                 // 图片文件
│   │   │   ├── ruter                  // 路由页面
│   │   │   ├── store                  // 数据管理
│   │   │   ├── style                  // 样式文件
│   │   │   ├── utils                  // 工具函数
```

```
|           |           ├── views              // 视图文件
|           |           ├── App.vue            // 应用
|           |           └── main.js            // 应用入口
|           |      └──index.html               // 前端 HTML 文件
|           ├── client                         // 前端站点
|           |       ├── src
|           |       |    ├── components        // 前端基础组件
|           |       |    ├── filters           // 工具函数
|           |       |    ├── images            // 图片文件
|           |       |    ├── ruter             // 路由页面
|           |       |    ├── store             // 数据管理
|           |       |    ├── style             // 样式文件
|           |       |    ├── utils             // 工具函数
|           |       |    ├── views             // 视图文件
|           |       |    ├── App.vue           // 应用
|           |       |    └── main.js           // 应用入口
|           |       └──index.html              // 前端 HTML 文件
|           └── server                         // 服务端
|                  ├── controller
|                  ├── middleware
|                  ├── models
|                  |    ├── blog.js            // blog 数据结构
|                  |    └── user.js            // user 数据结构
|                  ├── router
|                  ├── app.js                  // 启动应用
|                  ├── config.js               // 应用配置
|                  ├── index.js                // 入口文件
|                  └── mongodb.js              // 链接数据库
├── logs          // 日志文件
|     ├── all
|     └── info
├── node_modules                               // 项目依赖模块
├── publish
|     ├── static                               // 静态文件
|     └── index.html                           // HTML 页面
├── static                                     // 项目静态文件的入口
├── .babelrc
├── .gitignore
├── package.json
├── postcss.config.js
└── README.md
```

其中，build 文件夹存放 Webpack 的配置文件，code 中有 3 个文件夹，分别为 admin 后台管理、client 前端站点、server 服务端代码，config 中存放项目参数配置的文件，static 为项目静态文件的入口，node_modules 存放项目依赖模块，logs 存放日志文件，.babelrc 是 babel 编译配置，postcss.config.js 是 CSS 后处理器的配置。

12.2 数据库设计

12.2.1 数据准备

首先使用命令 use test 切换 MongoDB 数据库，创建 users 集合：

```
> mongo
> use test 切换数据库
switched to db test
> db.createCollection("users") 创建集合 users
> show collections 显示集合
users
```

创建管理员账户 admin：

```
> db.users.insert({
...     "name" : "test",
...     "pwd" : "e10adc3949ba59abbe56e057f20f883e",
...     "username" : "admin",
...     "roles" : [
...         "admin"
...     ]
... })
WriteResult({ "nInserted" : 1 })
> db.users.find().pretty()
```

插入成功之后通过 db.users.find().pretty() 查询文档。

12.2.2 连接数据库

引入 mongoose 连接数据库，示例使用 test 数据库，代码如下：

```
import mongoose from 'mongoose'
const DB_URL = 'mongodb://localhost:27017/test'

mongoose.Promise = global.Promise
mongoose.connect(DB_URL, { useMongoClient: true }, err => {
    if (err) {
        console.log("数据库连接失败！")
    }else{
        console.log("数据库连接成功！")
    }
})
export default mongoose
```

12.2.3 创建表结构

博客网站涉及两个数据表：blog 和 user。

（1）blog 表结构：

```
import db from '../mongodb'
let blogSchema = db.Schema({
  type: Array,
  title: String,
  desc: String,
  html: String,
  markdown: String,
  level: Number,
  github: String,
  source: Number,
  isVisible: Boolean,
  releaseTime: Date,
  createTime: {
    type: Date,
    default: Date.now
  }
})
export default db.model('blog', blogSchema)
```

（2）user 表结构：

```
import db from '../mongodb'
let userSchema = db.Schema({
  username: String,
  pwd: String,
  name: String,
  avatar: String,
  roles: Array,
  createTime: {
    type: Date,
    default: Date.now
  },
  loginTime: Date
})
export default db.model('user', userSchema);
```

12.3 服务端实现

12.3.1 启动后台应用

在 app.js 中引入 Koa，启动应用：

```
import Koa from 'koa'
import ip from 'ip'
import conf from './config'
import router from './router'
import middleware from './middleware'
import './mongodb'

const app = new Koa()
middleware(app)
router(app)
app.listen(conf.port, '0.0.0.0', () => {
    console.log(`server is running at http://${ip.address()}:${conf.port}`)
})
```

Koa 是基于 Node.js 平台的一个 Web 开发框架，由 Express 原班人马打造，致力于成为 Web 应用和 API 开发领域中更小且更加富有表现力、更健壮的基石。通过 async 函数 Koa 帮用户丢弃回调函数，并有力地增强错误处理。Koa 并没有绑定任何中间件，而是提供了一套优雅的方法帮助用户快速编写服务器端应用程序。

12.3.2 配置中间件

Koa 把很多 async 函数组成一个处理链，每个 async 函数都可以做一些自己的事情，然后用 await next() 来调用下一个 async 函数。这里把每个 async 函数称为 middleware，这些 middleware 可以组合起来，完成很多有用的功能。

Koa 通过 app.use(function)将给定的中间件方法添加到此应用程序，通过 app.use()注册的 async 函数会传入 ctx 和 next 参数，可以对 ctx 操作，并设置返回内容：

```
import path from 'path'
import bodyParser from 'koa-bodyparser'
import staticFiles from 'koa-static'
import Rule from './rule'
import Send from './send'
import Auth from './auth'
import Log from './log'
import Func from './func'
```

```javascript
export default app => {

    //缓存拦截器
    app.use(async (ctx, next) => {
        if (ctx.url == '/favicon.ico') return

        await next()
        ctx.status = 200
        ctx.set('Cache-Control', 'must-revalidation')
        if (ctx.fresh) {
            ctx.status = 304
            return
        }
    })

    // 日志中间件
    app.use(Log())
    // 数据返回的封装
    app.use(Send())
    // 方法封装
    app.use(Func())
    //权限中间件
    app.use(Auth())
    //post 请求中间件
    app.use(bodyParser())
    //静态文件中间件
    app.use(staticFiles(path.resolve(__dirname, '../../../public')));

    // 规则中间件
    Rule({
        app,
        rules: [{
                path: path.join(__dirname, '../controller/admin'),
                name: 'admin'
            },
            {
                path: path.join(__dirname, '../controller/client'),
                name: 'client'
            }
        ]
    })

    // 增加错误的监听处理
    app.on("error", (err, ctx) => {
        if (ctx && !ctx.headerSent && ctx.status < 500) {
```

```
            ctx.status = 500
        }
        if (ctx && ctx.log && ctx.log.error) {
            if (!ctx.state.logged) {
                ctx.log.error(err.stack)
            }
        }
    })
}
```

示例中注册了缓存中间件、日志中间件、权限中间件、post 请求中间件、静态文件中间件、规则中间件。调用 app.use() 的顺序决定了 middleware 的顺序。如果一个 middleware 没有调用 await next()，后续的 middleware 就不再执行。例如，若日志中间件中有调用 await next()，则会继续执行后续的中间件：

```
import log4js from 'log4js'
import access from './access' // 引入日志输出信息的封装文件
import config from '../../config'
const methods = ["trace", "debug", "info", "warn", "error", "fatal", "mark"];

// 提取默认公用参数对象
const baseInfo = config.log
export default (options = {}) => {
    let contextLogger = {},    //错误日志等级对象，最后会赋值给 ctx 上，用于打印各种日志
        appenders = {},        //日志配置
        opts = Object.assign({}, baseInfo, options), //系统配置
        {
            logLevel,
            dir,
            ip,
            projectName
        } = opts,
        commonInfo = {
            projectName,
            ip
        }; //存储公用的日志信息

    //指定要记录的日志分类
    appenders.all = {
        type: 'dateFile',                            //日志文件类型，可以使用日期作为文件名的占位符
        filename: '${dir}/all/',                     //日志文件名，可以设置相对路径或绝对路径
        pattern: 'task-yyyy-MM-dd.log',              //占位符，紧跟在 filename 后面
        alwaysIncludePattern: true                   //是否总是有后缀名
    }
```

```js
        // 环境变量为dev local development 认为是开发环境
    if (config.env === "dev" || config.env === "local" || config.env === "development") {
        appenders.out = {
            type: "console"
        }
    }

    let logConfig = {
        appenders,

        /**
         * 指定日志的默认配置项
         * 如果log4js.getLogger 中没有指定,就默认为 cheese 日志的配置项
         */
        categories: {
            default: {
                appenders: Object.keys(appenders),
                level: logLevel
            }
        }
    }

    let logger = log4js.getLogger('cheese');
    return async (ctx, next) => {
        const start = Date.now() // 记录请求开始的时间

        // 循环methods 将所有方法挂载到ctx 上
        methods.forEach((method, i) => {
            contextLogger[method] = message => {
                logConfig.appenders.cheese = {
                    type: 'dateFile',    //日志文件类型,可以使用日期作为文件名的占位符
                    filename: '${dir}/${method}/',
                    pattern: '${method}-yyyy-MM-dd.log',
                    alwaysIncludePattern: true        //是否总是有后缀名
                }
                log4js.configure(logConfig)
                logger[method](access(ctx, message, commonInfo))
            }
        })
        ctx.log = contextLogger
        await next()
        // 记录完成的时间作差,计算响应时间
        const responseTime = Date.now() - start
```

```
        ctx.log.info(access(ctx, {
            responseTime: '响应时间为${responseTime/1000}s'
        }, commonInfo))
    }
}
```

中间件函数是一个带有 ctx 和 next 两个参数的简单函数，next 用于把中间件的执行权限交给下游中间件，而当前中间件中，位于 next()之后的代码会暂停执行，直到最后一个中间件执行完毕，再自下而上依次执行每个中间件中 next()之后的代码，类似于一种先进后出的堆栈结构。

12.3.3 搭建路由和控制器

路由的工作是将具体的访问路径指向特定的功能模块，例如针对首页、列表页指向特定的功能模块的示例代码如下，未指定时返回 404：

```
const koa = require('koa2')
const app = new koa()

app.use(async (ctx, next) => {
    if (ctx.request.path === '/') {         // 首页
        ctx.response.status = 200
        ctx.response.body = 'index'
    } else if (ctx.request.path === '/list') {   // 列表页
        ctx.response.status = 200
        ctx.response.body = 'list'
    } else {
        ctx.throw(404, 'Not found')          // 404
    }
    await next()
})

app.listen(3000)
```

可以根据 ctx.request.path 增加新的路径判断。但是这种处理非常烦琐，Koa 社区提供了封装模块来解决这个问题，即 koa-router。

服务端的路由接口包括管理后台的接口和前端站点涉及的接口。通过路由中间件 koa-router 对路由进行控制：

```
import koaRouter from 'koa-router'
const router = koaRouter()

export default app => {
```

```
    /*----------------------admin--------------------------*/
    // 用户请求
    router.post('/admin_demo_api/user/login', app.admin.user.login)
    router.get('/admin_demo_api/user/info', app.admin.user.info)
    router.get('/admin_demo_api/user/list', app.admin.user.list)
    router.post('/admin_demo_api/user/add', app.admin.user.add)
    router.post('/admin_demo_api/user/update', app.admin.user.update)
    router.get('/admin_demo_api/user/del', app.admin.user.del)

    // 文章请求
    router.get('/admin_demo_api/blog/list', app.admin.blog.list)
    router.post('/admin_demo_api/blog/add', app.admin.blog.add)
    router.post('/admin_demo_api/blog/update', app.admin.blog.update)
    router.get('/admin_demo_api/blog/del', app.admin.blog.del)

    // 其他请求
    router.post('/admin_demo_api/markdown_upload_img',
app.admin.other.markdown_upload_img)


    /*----------------------client--------------------------*/
    // client/文章请求
    router.get('/client_demo_api/blog/list', app.client.blog.list)
    router.get('/client_demo_api/blog/info', app.client.blog.info)

    app.use(router.routes()).use(router.allowedMethods());
}
```

对应的 Controller 由规则中间件注册：

规则中间件实现：

```
import Path from 'path'
import fs from 'fs'

export default opts => {
    let {
        app,
        rules = []
    } = opts
    if (!app) {
        throw new Error("the app params is necessary!")
    }

    app.router = {};
```

```js
        const appKeys = Object.keys(app)
        rules.forEach((item) => {
            let {
                path,
                name
            } = item
            if (appKeys.includes(name)) {
                throw new Error(`the name of ${name} already exists!`)
            }
            let content = {};
            //readdirSync: 方法将返回一个包含"指定目录下所有文件名称"的数组对象
            //extname: 返回 path 路径文件扩展名，如果 path 以'.'结尾，就返回'.'，如果无扩展
名又不以'.'结尾，就返回空值
            //basename: path.basename(p, [ext]) p->要处理的 path ext->要过滤的字符
            fs.readdirSync(path).forEach(filename => {

                let extname = Path.extname(filename);
                if (extname === '.js') {
                    let name = Path.basename(filename, extname)
                    content[name] = require(Path.join(path, filename))
                    content[name].filename = name
                }
            })
            app[name] = content
        })
    }
```

注册规则中间件，将 controller 文件夹下的 admin 文件夹和 client 文件夹下的文件都注册为对应的 Controller：

```js
Rule({
    app,
    rules: [{
        path: path.join(__dirname, '../controller/admin'),
        name: 'admin'
    },
    {
        path: path.join(__dirname, '../controller/client'),
        name: 'client'
    }
    ]
})
```

controller 文件夹下的目录结构如下：

```
├── controller                       // 存放文件
│   ├── admin                        // 后台路由控制器
```

```
|       ├── blog.js              // 管理博客
|       ├── user.js              // 管理账户
|       └── other.js             // 管理其他
|   ├── client                   // 管理前台
|       └── blog.js              // 管理博客
```

12.3.4 账户管理

账户管理中包括增加账号、删除账号、查询账号列表、获取单个账号详细信息、更新账号信息等路由，实现代码如下：

```
// jwt 加密的使用
import jwt from 'jsonwebtoken'
import conf from '../../config'
import userModel from '../../models/user'
module.exports = {
    async login(ctx, next) {
        console.log('----------------登录接口 user/login----------------------');
        let {
            username,
            pwd
        } = ctx.request.body;
        console.log(username)
        try {
            let data = await ctx.findOne(userModel, {
                username: username
            });
            console.log(data)
            if (!data) {
                return ctx.sendError('用户名不存在！');
            }
            if (pwd !== data.pwd) {
                return ctx.sendError('密码错误,请重新输入！');
            }
            await ctx.update(userModel, {
                _id: data._id
            }, {
                $set: {
                    loginTime: new Date()
                }
            }) //更新登录时间

            let payload = {
                _id: data._id,
```

```javascript
                username: data.username,
                name: data.name,
                roles: data.roles
            }
            let token = jwt.sign(payload, conf.auth.admin_secret, {
                expiresIn: '24h'
            })                      //token 签名，有效期为24小时
            ctx.cookies.set(conf.auth.tokenKey, token, {
                httpOnly: false,    // 是否只用于 HTTP 请求中获取
            });
            console.log('登录成功')
            ctx.send({
                message: '登录成功'
            });
        } catch (e) {
            if (e === '暂无数据') {
                console.log('用户名不存在')
                return ctx.sendError('用户名不存在');
            }
            ctx.throw(e);
            ctx.sendError(e)
        }

    },
    async info(ctx, next) {
        console.log('----------------获取用户信息接口 user/getUserInfo----------------------');
        let token = ctx.request.query.token;
        try {
            let tokenInfo = jwt.verify(token, conf.auth.admin_secret);
            console.log(tokenInfo)
            ctx.send({
                username: tokenInfo.username,
                name: tokenInfo.name,
                _id: tokenInfo._id,
                roles: tokenInfo.roles
            })
        } catch (e) {
            if ('TokenExpiredError' === e.name) {
                ctx.sendError('鉴权失败,请重新登录!');
                ctx.throw(401, 'token expired,请及时本地保存数据！');
            }
            ctx.throw(401, 'invalid token');
            ctx.sendError('系统异常!');
        }
```

```javascript
    },

    async list(ctx, next) {
        console.log('-----------------获取用户信息列表接口 user/getUserList----------------------');
        let {
            keyword,
            pageindex = 1,
            pagesize = 10
        } = ctx.request.query;
        console.log('keyword:' + keyword + ',' + 'pageindex:' + pageindex + ',' + 'pagesize:' + pagesize)

        try {
            let reg = new RegExp(keyword, 'i')
            let data = await ctx.findPage(userModel, {
                $or: [{
                    name: {
                        $regex: reg
                    }
                },
                {
                    username: {
                        $regex: reg
                    }
                }
                ]
            }, {
                pwd: 0
            }, {
                limit: pagesize * 1,
                skip: (pageindex - 1) * pagesize
            });

            ctx.send(data)
        } catch (e) {
            console.log(e)
            ctx.sendError(e)
        }

    },

    async add(ctx, next) {
        console.log('-----------------添加管理员
```

```js
user/add---------------------');
        let paramsData = ctx.request.body;
        try {
            let data = await ctx.findOne(userModel, {
                name: paramsData.name
            })
            if (data) {
                ctx.sendError('数据已经存在，请重新添加!')
            } else {
                let data = await ctx.add(userModel, paramsData);
                ctx.send(paramsData)
            }
        } catch (e) {
            ctx.sendError(e)
        }
    },

    async update(ctx, next) {
        console.log('----------------更新管理员
user/update---------------------');
        let paramsData = ctx.request.body;
        console.log(paramsData)
        try {
            let data = await ctx.findOne(userModel, {
                name: paramsData.name
            })
            if (paramsData.old_pwd !== data.pwd) {
                return ctx.sendError('密码不匹配!')
            }
            delete paramsData.old_pwd
            await ctx.update(userModel, {
                _id: paramsData._id
            }, paramsData)
            ctx.send()
        } catch (e) {
            if (e === '暂无数据') {
                ctx.sendError(e)
            }
        }
    },

    async del(ctx, next) {
        console.log('----------------删除管理员
user/del---------------------');
        let id = ctx.request.query.id
```

```
        try {
           ctx.remove(userModel, {
              _id: id
           })
           ctx.send()
        } catch (e) {
           ctx.sendError(e)
        }
     }
  }
```

其中，jsonwebtoken 模块基于 JWT（JSON Web Token）规范生成账户认证机制。JWT 是一个非常轻巧的规范。该规范允许开发者使用 JWT 在用户和服务器之间传递安全可靠的信息。

JWT 的原理是，服务器认证用户成功之后生成一个 JSON 对象返回给客户端。之后客户端与服务端通信时，都需要返回该 JSON 对象。服务器完全通过该 JSON 对象识别用户身份。为了防止用户篡改数据，服务器在生成该 JOSN 对象时会加上签名。服务器就不需要再保存任何 session 数据了，也就是说服务器变成无状态，从而比较容易实现扩展。

12.3.5 博客管理

博客管理中涉及的路由包括添加博客、删除博客、查询博客列表、获取单个博客详细信息、更新博客等路由，实现代码如下：

```
import blogModel from '../../models/blog'
import path from 'path'
import marked from 'marked'

marked.setOptions({
   renderer: new marked.Renderer(),
   renderer: new marked.Renderer(),
   gfm: true,          //允许 GitHub 标准的 markdown
   tables: true,       //允许支持表格语法。该选项要求 gfm 为 true
   breaks: true,       //允许回车换行。该选项要求 gfm 为 true
   pedantic: false,    //尽可能地兼容 markdown.pl 的晦涩部分。不纠正原始模型任何的不良行为和错误
   sanitize: true,     //对输出进行过滤（清理），将忽略任何已经输入的 HTML 代码（标签）
   smartLists: true,   //使用比原生 markdown 更时髦的列表。旧的列表将可能被作为 pedantic 的处理内容过滤掉
   smartypants: false, //使用更为时髦的标点，比如在引用语法中加入破折号
   highlight: function (code) {
      return require('highlight.js').highlightAuto(code).value;
   }
})
```

```js
module.exports = {
    async list(ctx, next) {
        console.log('----------------获取博客列表 blog/list----------------------');
        let {
            keyword,
            pageindex = 1,
            pagesize = 10
        } = ctx.request.query;
        console.log('keyword:' + keyword + ',' + 'pageindex:' + pageindex + ',' + 'pagesize:' + pagesize)
        try {

            let reg = new RegExp(keyword, 'i')
            let data = await ctx.findPage(blogModel, {
                $or: [{
                    type: {
                        $regex: reg
                    }
                },
                {
                    title: {
                        $regex: reg
                    }
                }
                ]
            }, {
                createTime: 0,
                html: 0
            }, {
                limit: pagesize * 1,
                skip: (pageindex - 1) * pagesize
            });
            ctx.send(data)
        } catch (e) {
            console.log(e)
            ctx.sendError(e)
        }

    },

    async add(ctx, next) {
        console.log('----------------添加博客 blog/add----------------------');
```

```
        let paramsData = ctx.request.body;
        try {
            let data = await ctx.findOne(blogModel, {
                title: paramsData.title
            })
            if (data) {
                ctx.sendError('数据已经存在,请重新添加!')
            } else {


                paramsData.html = marked(paramsData.html);
                let data = await ctx.add(blogModel, paramsData);
                ctx.send(paramsData)
            }
        } catch (e) {
            ctx.sendError(e)
        }
    },

    async update(ctx, next) {
        console.log('----------------更新博客 blog/update----------------------');
        let paramsData = ctx.request.body;
        try {
            paramsData.html = marked(paramsData.html);
            let data = await ctx.update(blogModel, {
                _id: paramsData._id
            }, paramsData)
            ctx.send()
        } catch (e) {
            if (e === '暂无数据') {
                ctx.sendError(e)
            }
        }
    },

    async del(ctx, next) {
        console.log('----------------删除博客 blog/del----------------------');
        let id = ctx.request.query.id
        try {
            ctx.remove(blogModel, {
                _id: id
            })
            ctx.send()
```

```
        } catch (e) {
            ctx.sendError(e)
        }
    }
}
```

其中，marked 是一个用 JavaScript 编写的功能齐全的 Markdown 解析器和编译器，可以非常方便地在线编译 Markdown 代码为 HTML 并直接显示，并且支持完全地自定义各种格式，便于发布博客。

12.4 博客后台管理的实现

12.4.1 目录结构

博客后台管理系统的目录结构如下：

```
|   ├── admin              // 管理后台
|   |   ├── src
|   |   |   ├── components    // 前端基础组件
|   |   |   ├── filters       // 工具函数
|   |   |   ├── images        // 图片文件
|   |   |   ├── ruter         // 路由页面
|   |   |   ├── store         // 数据管理
|   |   |   ├── style         // 样式文件
|   |   |   ├── utils         // 工具函数
|   |   |   ├── views         // 视图文件
|   |   |   ├── App.vue       // 应用
|   |   |   └── main.js       // 应用入口
|   |   └──index.html         // 前端 HTML 文件
```

12.4.2 权限管理

后台权限管理的部分包括添加管理员、账号列表、账号信息的修改和删除账号等操作管理。添加管理员的示例代码如下：

```
<template>
  <article>
    <h2>添加管理员</h2>
    <div class="box">
      <el-form :model="info" :rules="rules" ref="form" label-width="100px" class="form">
        <el-form-item label="名字" prop="name">
```

```
            <el-input type="text" v-model="info.name"></el-input>
          </el-form-item>
          <el-form-item label="用户名" prop="username">
            <el-input type="text" v-model="info.username"></el-input>
          </el-form-item>
          <el-form-item label="密码" prop="pwd">
            <el-input type="password" v-model="info.pwd"></el-input>
          </el-form-item>
          <el-form-item label="权限" prop="roles">
            <el-select v-model="info.roles" multiple placeholder="请选择" class="block">
              <el-option
                v-for="item in roles"
                :key="item.value"
                :label="item.label"
                :value="item.value"
              ></el-option>
            </el-select>
          </el-form-item>
          <el-form-item>
            <el-button type="primary" @click="submitForm('form')" :loading="loading">立即创建</el-button>
          </el-form-item>
        </el-form>
      </div>
    </article>
  </template>

  <script>
  import { mapGetters } from "vuex";
  export default {
    data() {
      return {
        info: {
          name: "",
          username: "",
          pwd: "",
          avatar: "",
          roles: ["default"]
        },
        roles: [
          { label: "超级管理员", value: "admin" },
          { label: "普通管理员", value: "default" }
        ],
        loading: false,
```

```
      rules: {
        name: [{ required: true, message: "请填写名字", trigger: "blur" }],
        username: [
          { required: true, message: "请填写用户名", trigger: "blur" }
        ],
        pwd: [{ required: true, message: "请填写密码", trigger: "blur" }],
        roles: [
          {
            required: true,
            message: "请选择权限",
            trigger: "change",
            type: "array"
          }
        ]
      }
    };
  },
  methods: {
    submitForm(formName) {
      this.loading = true;
      this.$refs[formName].validate(async valid => {
        if (valid) {
          try {
            await this.$store.dispatch("addUser", this.info);
            this.loading = false;
            this.$router.push("/permission/list");
          } catch (e) {
            this.loading = false;
          }
        } else {
          console.log("error submit!!");
          this.loading = false;
          return false;
        }
      });
    }
  }
};
</script>


<style lang="less" scoped>
article {
  text-align: center;
  padding: 0 100px;
```

```
    h2 {
      text-align: center;
      line-height: 80px;
      color: #666;
    }
    .box {
      width: 500px;
      text-align: left;
    }
    .block {
      width: 100%;
      display: block;
    }
    .left-item {
      text-align: left;
    }
    .submit {
      width: 100px;
    }
  }
</style>
```

运行命令 npm run dev:client，打开浏览器，输入 http://localhost:8090/#/permission/add，显示的效果如图 12.1 所示。

图 12.1　添加管理员

管理员列表前端代码示例：

```
<template>
  <article>
    <div class="search">
      <el-input
        placeholder="请输入内容"
        prefix-icon="el-icon-search"
```

```html
          v-model="keyword"
          @keydown.enter.native="getUserList"
        ></el-input>
        <el-button type="primary" icon="el-icon-search" :loading="loading" @click="getUserList">搜索</el-button>
      </div>
      <el-table ref="multipleTable" :data="userList" tooltip-effect="dark" stripe border>
        <el-table-column
          type="index"
          width="55"
          align="center"
          header-align="center"
          :index="increment"
        ></el-table-column>

        <el-table-column
          show-overflow-tooltip
          v-if="!item.hidden"
          v-for="(item, index) in headerOptions"
          :key="index"
          :label="item.label"
          :prop="item.prop"
          :header-align="item.headerAlign"
          :align="item.align"
          :min-width="item.minWidth || 150"
        >
          <template slot-scope="scope">
            <div v-if="scope.column.property === 'roles'">
              <el-tag
                class="tag"
                type="primary"
                close-transition
                v-for="(tag, index) in scope.row.roles"
                :key="index"
              >{{tag}}</el-tag>
            </div>
            <div v-else-if="scope.column.property === 'avatar'">
              <img :src="scope.row[scope.column.property]" alt style="height: 40px;" />
            </div>
            <div v-else>{{scope.row[scope.column.property] || '无'}}</div>
          </template>
        </el-table-column>
        <el-table-column label="操作" header-align="center" align="center" width="250">
          <template slot-scope="scope">
```

```html
              <el-button size="mini" @click="edit(scope)">编辑</el-button>
              <el-button size="mini" type="danger" @click="del(scope)">删除</el-button>
            </template>
          </el-table-column>
      </el-table>
      <el-pagination
        class="pagination"
        @size-change="handleSizeChange"
        @current-change="handleCurrentChange"
        :current-page="pageindex"
        :page-sizes="size_scoped"
        :page-size="pagesize"
        layout="total, sizes, prev, pager, next, jumper"
        :total="userTotal"
      ></el-pagination>
      <EditComponent v-if="editShow" :info="userInfo" @close="close"></EditComponent>
    </article>
  </template>
  <script>
  import { mapGetters } from "vuex";
  import EditComponent from "../edit/index";
  export default {
    components: {
      EditComponent
    },
    data() {
      return {
        keyword: "",
        editShow: false,
        userInfo: {},
        loading: false,
        pageindex: 1,
        pagesize: 10,
        size_scoped: [10, 20, 30, 40],
        headerOptions: [
          {
            label: "id",
            prop: "_id",
            hidden: true,
            headerAlign: "center",
            align: "center",
            width: ""
          },
          {
            label: "用户名",
```

```
        prop: "username",
        hidden: false,
        headerAlign: "center",
        align: "center",
        width: ""
      },
      {
        label: "姓名",
        prop: "name",
        hidden: false,
        headerAlign: "center",
        align: "center",
        width: "",
        sort: true
      },
      {
        label: "权限",
        prop: "roles",
        hidden: false,
        headerAlign: "center",
        align: "center",
        width: "",
        sort: true
      }
    ],
    multipleSelection: []
  };
},
mounted() {
  this.getUserList();
},

methods: {
  increment(index) {
    return index + 1 + (this.pageindex - 1) * this.pagesize;
  },
  close() {
    this.editShow = false;
    this.getUserList();
  },
  handleSizeChange(val) {
    // console.log(`每页 ${val} 条`);
    this.pagesize = val;
    this.getUserList();
  },
  handleCurrentChange(val) {
    // console.log(`当前页: ${val}`);
```

```js
      this.pageindex = val;
      this.getUserList();
    },
    async getUserList() {
      this.loading = true;
      try {
        await this.$store.dispatch("getUserList", {
          keyword: this.keyword,
          pageindex: this.pageindex,
          pagesize: this.pagesize
        });
        this.loading = false;
      } catch (e) {
        this.loading = false;
      }
    },
    del(scope) {
      this.$confirm("此操作将永久删除该文件, 是否继续?", "提示", {
        confirmButtonText: "确定",
        cancelButtonText: "取消",
        type: "warning",
        center: true
      })
        .then(async () => {
          try {
            await this.$store.dispatch("delUser", scope.row._id);
            this.userList.splice(scope.$index, 1);
          } catch (e) {}
          this.$message({
            type: "success",
            message: "删除成功!"
          });
        })
        .catch(() => {
          this.$message({
            type: "info",
            message: "已取消删除"
          });
        });
    },
    edit(scope) {
      console.log(scope);
      this.editShow = true;
      scope.row.releaseTime = new Date(scope.row.releaseTime);
      this.userInfo = scope.row;
    },
    filterTag(value, row) {
```

```
          return row.type.some(v => v === value);
        }
      },
      computed: {
        ...mapGetters(["userList", "userTotal"])
      }
    };
</script>

<style lang="less" scoped>
article {
  padding: 20px;
  .search {
    padding-bottom: 20px;
    .el-input {
      width: 300px;
    }
  }
  .pagination {
    text-align: right;
    padding: 20px 0;
  }
  .tag {
    margin: 0 10px;
  }
}
</style>
```

在浏览器中打开 http://localhost:8090/#/permission/list，显示如图 12.2 所示。

图 12.2　管理员列表

其中，调用服务端用户信息接口处理的代码如下：

```
import axios from 'src/utils/fetch'
import {
  getToken
} from 'src/utils/auth'
```

```js
import md5 from 'js-md5'

const user = {
  state: {
    list: [],
    userTotal: 0,
    name: '',
    username: '',
    roles: null,
    token: getToken(),
    otherList: []
  },
  mutations: {
    SET_TOKEN(state, token) {
      state.token = token;
    },
    SET_USERINFO(state, info) {
      state.name = info.name;
      state.username = info.username;
      state.roles = info.roles;
    },
    USERLIST(state, data) {
      state.list = data.list
      state.total = data.total;
    },
    GET_INFOLIST(state, data) {
      state.otherList = data;
    },
    CLEARINFO(state) {
      state.name = '';
      state.username = '';
      state.roles = null;
    }
  },
  actions: {
    clearInfo({
      commit
    }) {
      commit('CLEARINFO')
    },
    userLogin({
      state,
      commit
    }, info) {
      let {
```

```js
      username,
      pwd
    } = info;
    return new Promise((resolve, reject) => {
      axios.post('user/login', {
        username: username,
        pwd: md5(pwd)
      }).then(res => {
        // console.log(res)
        state.token = getToken();
        resolve(res)
      }).catch(err => {
        // console.log(err)
        reject(err)
      })
    })
  },
  getUserInfo({
    state,
    commit
  }) {
    return new Promise((resolve, reject) => {
      axios.get('user/info', {
        token: state.token
      }).then(res => {
        console.log(res)
        commit('SET_USERINFO', res.data)
        resolve(res)
      }).catch(err => {
        // console.log(err)
        reject(err)
      })
    })
  },
  getUserList({
    commit
  }, params) {
    return new Promise((resolve, reject) => {
      axios.get('user/list', params).then(res => {
        console.log(res)
        commit('USERLIST', res.data)
        resolve(res)
      }).catch(err => {
        // console.log(err)
        reject(err)
```

```
      })
    })
  },
  addUser({
    commit
  }, info) {
    info.pwd = md5(info.pwd)
    return new Promise((resolve, reject) => {
      axios.post('user/add', info)
        .then(res => {
          resolve(res)
        }).catch(err => {
          reject(err)
        })
    })
  },
  delUser({
    commit
  }, id) {
    return new Promise((resolve, reject) => {
      axios.get('user/del', {
          id: id
        })
        .then(res => {
          resolve(res)
        }).catch(err => {
          reject(err)
        })
    })
  },
  updateUser({
    commit
  }, info) {
    info.pwd = md5(info.pwd)
    info.old_pwd = md5(info.old_pwd)
    return new Promise((resolve, reject) => {
      axios.post('user/update', info)
        .then(res => {
          resolve(res)
        }).catch(err => {
          reject(err)
        })
    })
  }
}
```

```
}
export default user
```

打开浏览器控制台，可以看到对应的接口调用，如图 12.3 和图 12.4 所示。

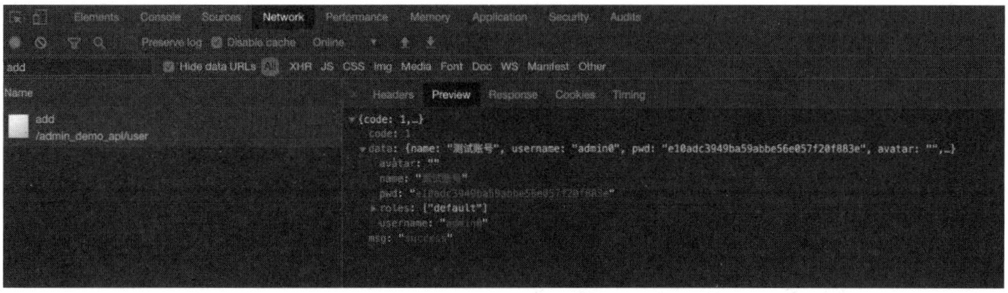

图 12.3 添加管理员

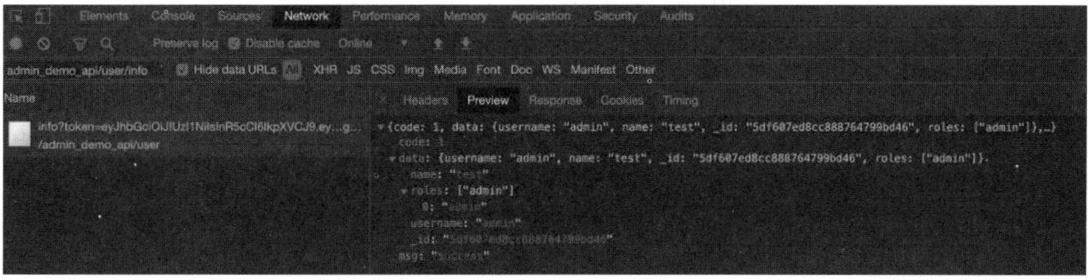

图 12.4 查询账户接口

12.4.3 博客管理

博客后台管理的部分包括添加博客、博客列表、对博客进行修改和删除已有博客等操作管理。添加博客的示例代码如下：

```
<template>
  <article>
    <h2>添加博客</h2>
    <div class="box">
      <el-form :model="info" :rules="rules" ref="form" label-width="100px" class="form">
        <el-form-item label="博客类型" prop="type">
          <el-select v-model="info.type" multiple clearable placeholder="请选择博客类型" class="block">
            <el-option
              v-for="item in blogTypes"
              :key="item.name"
              :label="item.name"
              :value="item.name"
            ></el-option>
          </el-select>
```

```html
        </el-form-item>
        <el-form-item label="文章标题" prop="title">
          <el-input type="text" v-model="info.title"></el-input>
        </el-form-item>
        <el-form-item label="文章描述" prop="desc">
          <el-input type="textarea" v-model="info.desc"></el-input>
        </el-form-item>
        <el-form-item label="文章内容" prop="markdown">
          <Markdown v-model="info.markdown"></Markdown>
        </el-form-item>
        <el-form-item label="级别" prop="album">
          <el-select v-model="info.level" placeholder="请选择级别" class="block">
            <el-option v-for="item in [1,2,3,4,5,6]" :key="item" :label="item" :value="item"></el-option>
          </el-select>
        </el-form-item>
        <el-form-item label="来源" prop="source">
          <el-select v-model="info.source" placeholder="请选择来源" class="block">
            <el-option v-for="item in sources" :key="item.id" :label="item.name" :value="item.id"></el-option>
          </el-select>
        </el-form-item>
        <el-form-item label="Github" prop="github">
          <el-input type="text" v-model="info.github"></el-input>
        </el-form-item>
        <el-form-item label="发布时间" prop="releaseTime">
          <el-date-picker class="block" v-model="info.releaseTime" type="date" placeholder="选择发布日期"></el-date-picker>
        </el-form-item>
        <el-form-item label="是否可见" prop="isVisible" class="left-item">
          <el-switch v-model="info.isVisible"></el-switch>
        </el-form-item>
        <el-form-item>
          <el-button type="primary" @click="submitForm('form')" :loading="loading">立即创建</el-button>
        </el-form-item>
      </el-form>
    </div>
  </article>
</template>

<script>
import { mapGetters } from "vuex";
```

```js
import Markdown from "components/Markdown";
export default {
  components: { Markdown },
  data() {
    return {
      info: {
        type: ["JavaScript"],
        title: "",
        desc: "",
        html: "",
        markdown: "",
        level: 1,
        source: 1,
        github: "",
        isVisible: true,
        releaseTime: new Date()
      },
      loading: false,
      rules: {
        type: [
          {
            required: true,
            message: "请至少选择一个文章类型",
            trigger: "change",
            type: "array"
          }
        ],
        title: [{ required: true, message: "请填写文章标题", trigger: "blur" }],
        desc: [{ required: true, message: "请填写文章描述", trigger: "blur" }],
        isVisible: [
          {
            required: true,
            message: "请选择",
            trigger: "change",
            type: "boolean"
          }
        ],
        releaseTime: [
          {
            required: true,
            message: "请选择文章的发布时间",
            trigger: "change",
            type: "date"
          }
        ]
```

```js
      }
    };
  },
  methods: {
    submitForm(formName) {
      this.loading = true;
      if (!this.info.markdown) {
        this.$message.warn("请填写文章内容");
        return;
      }
      this.$refs[formName].validate(async valid => {
        if (valid) {
          try {
            this.info.html = this.info.markdown;
            await this.$store.dispatch("addBlog", this.info);
            this.loading = false;
            this.$router.push("/article/list");
          } catch (e) {
            this.loading = false;
          }
        } else {
          console.log("error submit!!");
          this.loading = false;
          return false;
        }
      });
    }
  },
  computed: {
    ...mapGetters(["blogTypes", "sources"])
  }
};
</script>

<style lang="less" scoped>
article {
  text-align: center;
  padding: 0 100px;
  h2 {
    text-align: center;
    line-height: 80px;
    color: #666;
  }
  .block {
    width: 100%;
```

```
    display: block;
  }
  .left-item {
    text-align: left;
  }
  .submit {
    width: 100px;
  }
}
</style>
```

运行命令 npm run dev:client，打开浏览器，输入 http://localhost:8090/#/article/add，显示的效果如图 12.5 所示。

图 12.5　添加博客

文章列表的前端代码如下：

```
<template>
  <article>
```

```html
      <div class="search">
        <el-input
          placeholder="请输入内容"
          prefix-icon="el-icon-search"
          v-model="keyword"
          @keydown.enter.native="getBlogList"
        ></el-input>
        <el-button type="primary" icon="el-icon-search" :loading="loading" @click="getBlogList">搜索</el-button>
      </div>
      <el-table ref="multipleTable" :data="blogList" tooltip-effect="dark" stripe border>
        <el-table-column
          type="index"
          width="55"
          align="center"
          header-align="center"
          :index="increment"
        ></el-table-column>

        <el-table-column
          show-overflow-tooltip
          v-if="!item.hidden && !item.filters"
          v-for="(item, index) in headerOptions"
          :key="index"
          :label="item.label"
          :prop="item.prop"
          :header-align="item.headerAlign"
          :align="item.align"
          :sortable="item.sort"
          :min-width="item.minWidth || 100"
        >
          <template slot-scope="scope">
            <div
              v-if="scope.column.property == 'isVisible'"
            >{{scope.row[scope.column.property]?'是':'否'}}</div>
            <div
              v-else-if="scope.column.property == 'source'"
            >{{scope.row[scope.column.property] === 1?'原创':scope.row[scope.column.property] === 2?'转载':'翻译'}}</div>
            <div
              v-else-if="scope.column.property == 'releaseTime'"
            >{{scope.row[scope.column.property] | parseTime('{y}-{m}-{d}')}}</div>
            <div v-else>{{scope.row[scope.column.property] || '无'}}</div>
```

```html
        </template>
      </el-table-column>
      <el-table-column
        show-overflow-tooltip
        v-else-if="!item.hidden && item.filters"
        :key="index"
        :label="item.label"
        :prop="item.prop"
        :header-align="item.headerAlign"
        :align="item.align"
        :sortable="item.sort"
        :filters="item.filters"
        :filter-method="filterTag"
        :min-width="item.minWidth || 200"
      >
        <template slot-scope="scope">
          <el-tag
            class="tag"
            type="primary"
            close-transition
            v-for="(tag, index) in scope.row.type"
            :key="index"
          >{{tag}}</el-tag>
        </template>
      </el-table-column>
      <el-table-column label="操作" header-align="center" align="center" width="200">
        <template slot-scope="scope">
          <el-button size="mini" @click="edit(scope)">编辑</el-button>
          <el-button size="mini" type="danger" @click="del(scope)">删除</el-button>
        </template>
      </el-table-column>
    </el-table>
    <el-pagination
      class="pagination"
      @size-change="handleSizeChange"
      @current-change="handleCurrentChange"
      :current-page="pageindex"
      :page-sizes="size_scoped"
      :page-size="pagesize"
      layout="total, sizes, prev, pager, next, jumper"
      :total="blogTotal"
    ></el-pagination>
    <EditComponent v-if="editShow" :info="blogInfo"
```

```
@close="close"></EditComponent>
    </article>
  </template>
  <script>
  import { mapGetters } from "vuex";
  import EditComponent from "../edit/index";
  import { blogFilters } from "store/modules/classify";
  export default {
    components: {
      EditComponent
    },
    data() {
      return {
        keyword: "",
        editShow: false,
        blogInfo: {},
        loading: false,
        pageindex: 1,
        pagesize: 10,
        size_scoped: [10, 20, 30, 40],
        headerOptions: [
          {
            label: "_id",
            prop: "_id",
            hidden: true,
            headerAlign: "center",
            align: "center",
            width: ""
          },
          {
            label: "类型",
            prop: "type",
            hidden: false,
            headerAlign: "center",
            align: "center",
            width: "",
            filters: blogFilters
          },
          {
            label: "标题",
            prop: "title",
            hidden: false,
            headerAlign: "center",
            align: "center",
            width: "",
```

```
      sort: true
    },
    {
      label: "描述",
      prop: "desc",
      hidden: false,
      headerAlign: "center",
      align: "center",
      width: ""
    },
    {
      label: "来源",
      prop: "source",
      hidden: false,
      headerAlign: "center",
      align: "center",
      width: 80
    },
    {
      label: "级别",
      prop: "level",
      hidden: false,
      headerAlign: "center",
      align: "center",
      width: 30
    },
    {
      label: "发布时间",
      prop: "releaseTime",
      hidden: false,
      headerAlign: "center",
      align: "center",
      width: 100,
      sort: true
    },
    {
      label: "是否可见",
      prop: "isVisible",
      hidden: false,
      headerAlign: "center",
      align: "center",
      minWidth: 80
    }
  ],
  multipleSelection: []
```

```js
    };
  },
  mounted() {
    this.getBlogList();
  },

  methods: {
    increment(index) {
      return index + 1 + (this.pageindex - 1) * this.pagesize;
    },
    close() {
      this.editShow = false;
      this.getBlogList();
    },
    handleSizeChange(val) {
      // console.log(`每页 ${val} 条`);
      this.pagesize = val;
      this.getBlogList();
    },
    handleCurrentChange(val) {
      // console.log(`当前页: ${val}`);
      this.pageindex = val;
      this.getBlogList();
    },
    async getBlogList() {
      this.loading = true;
      try {
        await this.$store.dispatch("getBlogList", {
          keyword: this.keyword,
          pageindex: this.pageindex,
          pagesize: this.pagesize
        });
        this.loading = false;
      } catch (e) {
        this.loading = false;
      }
    },
    del(scope) {
      this.$confirm("此操作将永久删除该文件,是否继续?", "提示", {
        confirmButtonText: "确定",
        cancelButtonText: "取消",
        type: "warning",
        center: true
      })
        .then(async () => {
```

```js
          try {
            await this.$store.dispatch("delBlog", scope.row._id);
            this.blogList.splice(scope.$index, 1);
          } catch (e) {}
          this.$message({
            type: "success",
            message: "删除成功!"
          });
        })
        .catch(() => {
          this.$message({
            type: "info",
            message: "已取消删除"
          });
        });
    },
    edit(scope) {
      console.log(scope);
      this.editShow = true;
      scope.row.releaseTime = new Date(scope.row.releaseTime);
      this.blogInfo = scope.row;
    },
    filterTag(value, row) {
      return row.type.some(v => v === value);
    }
  },
  computed: {
    ...mapGetters(["blogList", "blogTotal"])
  }
};
</script>

<style lang="less" scoped>
article {
  padding: 20px;
  .search {
    padding-bottom: 20px;
    .el-input {
      width: 300px;
    }
  }
  .pagination {
    text-align: right;
    padding: 20px 0;
  }
```

```css
    .tag {
      margin: 0 10px;
    }
}
</style>
```

在浏览器中打开 http://localhost:8090/#/article/list，文章列表页展示效果如图 12.6 所示。

图 12.6　文章列表页

博客管理中调用服务端接口的处理代码如下，包括添加博客方法 addBlog、获取博客列表方法 getBlogList、删除博客方法 delBlog、更新博客方法 updateBlog：

```js
import axios from 'src/utils/fetch'
import { blogTypes } from './classify'

const music = {
  state: {
    blogTypes,
    list: [],
    total: 0
  },
  mutations: {
    BLOGLIST (state, data) {
      state.list = data.data.list;
      state.total = data.data.total;
    }
  },
  actions: {
    addBlog ({commit}, info) {
      return new Promise( (resolve, reject) => {
        axios.postFile('blog/add', info)
          .then( res => {
            resolve(res)
          }).catch( err => {
            reject(err)
          })
      })
```

```
    },
    getBlogList ({commit}, params) {
      return new Promise( (resolve, reject) => {
        axios.get('blog/list', params)
          .then( res => {
            commit('BLOGLIST', res)
            resolve(res)
          }).catch( err => {
            reject(err)
          })
      })
    },
    delBlog ({commit}, id) {
      return new Promise( (resolve, reject) => {
        axios.get('blog/del', {id: id})
          .then( res => {
            resolve(res)
          }).catch( err => {
            reject(err)
          })
      })
    },
    updateBlog ({commit}, info) {
      return new Promise( (resolve, reject) => {
        axios.postFile('blog/update', info)
          .then( res => {
            resolve(res)
          }).catch( err => {
            reject(err)
          })
      })
    }
  }
}
export default music
```

添加博客时,打开浏览器控制台,可以看到对应的添加博客接口和博客列表接口的查询请求,如图12.7和图12.8所示。

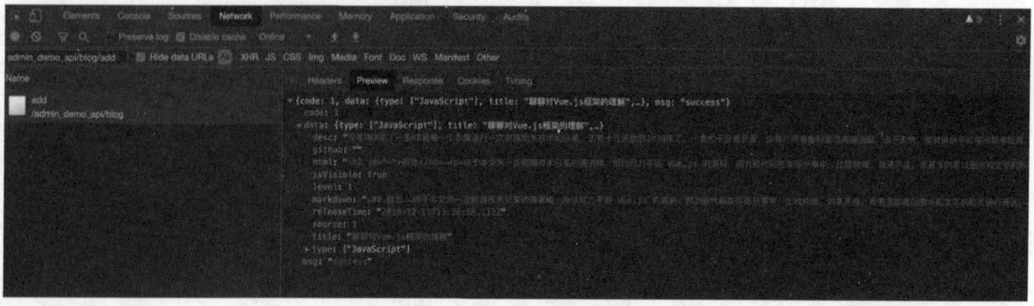

图 12.7　添加博客接口

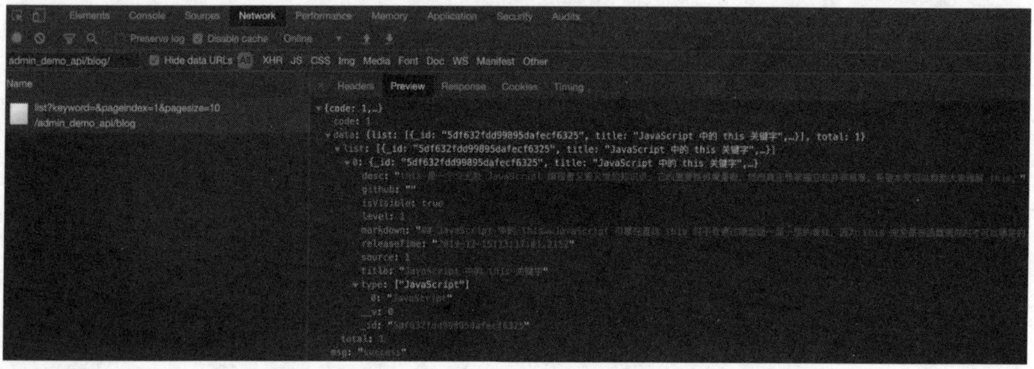

图 12.8　博客列表接口

12.5　博客前台站点的实现

12.5.1　目录结构

博客前台站点的目录结构如下：

```
|   ├── client              // 前端站点
|   │   ├── src
|   │   │   ├── components      // 前端基础组件
|   │   │   │   ├── Back            // 返回组件
|   │   │   │   ├── Github          // Github 组件
|   │   │   │   ├── Icon-svg        // Svg 的 Icon 组件
|   │   │   │   ├── Loading         // Loading 组件
|   │   │   │   ├── Markdown        // Markdown 组件
|   │   │   │   ├── NoneData        // 占位组件
|   │   │   │   └── Tag             // Tag 组件
|   │   │   ├── filters         // 工具函数
|   │   │   ├── images          // 图片文件
|   │   │   ├── ruter           // 路由页面
|   │   │   │   └── index.js
```

```
|           ├── store              // 数据管理
|           |    └── index.js
|           ├── style              // 样式文件
|           ├── utils              // 工具函数
|           |    ├── auth
|           |    ├── storage
|           |    └── fetch.js
|           ├── views              // 视图文件
|           |    ├── Artcle
|           |    ├── home
|           |    |    ├── blog.vue
|           |    |    ├── index.vue
|           |    |    ├── info.vue
|           |    |    └── tags.vue
|           ├── App.vue            // 应用
|           └── main.js            // 应用入口
└──index.html                      // 前端 HTML 文件
```

12.5.2 博客列表页

博客列表页首先展示标签分类，根据博客的标签分类筛选之后，可以展示对应类型的博客列表，代码示例如下：

```
<template>
  <div class="home-wrapper cf">
    <infoComponent></infoComponent>
    <div :class="{'tags-box': pc_bol}">
      <TagsComponent></TagsComponent>
    </div>
    <div :class="{'view-box': pc_bol}">
      <router-view>
        <BlogComponent></BlogComponent>
      </router-view>
    </div>
  </div>
</template>
<script>
import { mapGetters } from "vuex";
import infoComponent from "./info.vue";
import BlogComponent from "./blog.vue";
import TagsComponent from "./tags.vue";
export default {
  data() {
    return {
      winH: document.documentElement.clientHeight ||
```

```
      document.body.clientHeight
        };
    },
    components: {
      infoComponent,
      BlogComponent,
      TagsComponent
    },
    mounted() {
      window.addEventListener("scroll", () => {
        let distance =
          document.documentElement.scrollTop || document.body.scrollTop,
         scrollH =
          document.documentElement.scrollHeight || document.body.scrollHeight;

        if (distance + this.winH >= scrollH) {
          if (this.blogLoadingBol) {
            console.log(111);
            // this.pageindex ++;
            // this.$store.dispatch('getBlogList', {
            //    type: this.$route.params.classify,
            //    pageindex: this.pageindex
            // })
          }
        }
      });
    },
    computed: {
      ...mapGetters(["pc_bol"])
    }
};
</script>
<style lang="less" scoped>
.tags-box {
  width: 30%;
  float: left;
}
.view-box {
  width: 67%;
  float: right;
}
</style>
```

运行命令 npm run dev:client,在浏览器中打开 http://localhost:8080/blog/JavaScript,显示标记为 JavaScript 类的博客列表,如图 12.9 所示。

第 12 章 实战：基于 Koa+MongoDB 实现博客网站

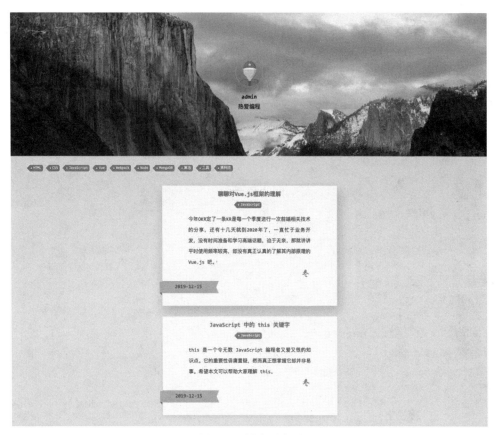

图 12.9 博客列表页

处理博客列表的接口交互代码示例：

```
import axios from '../../utils/fetch'
import {
    blogTypes
} from './classify'

const blog = {
    state: {
        types: blogTypes,
        list: [],
        homeList: [],
        info: {},
        currType: '',
        pagesize: 5,
        loadingMore: false,
        loadingBol: true
    },
    mutations: {
        BLOGLIST(state, res) {
```

```js
                state.list = res;
            },
            BLOGINFO(state, res) {
                state.info = res.data;
            }
        },
        actions: {
            // 获取博客列表
            async getBlogList({
                commit,
                state
            }, params) {

                params.pagesize = params.pagesize || state.pagesize
                params.type = params.type === 'all' ? null : params.type
                state.loadingMore = true
                state.loadingBol = false
                return new Promise((resolve, reject) => {
                    axios.get('blog/list', params).
                    then(res => {
                        state.loadingMore = false;
                        resolve(res)
                        if (res.data.length <= 0 && params.pageindex > 1) return
                        if (params.pageindex > 1) {
                            commit('BLOGLIST', state.list.concat(res.data))
                        } else {
                            commit('BLOGLIST', res.data)
                        }
                        if (res.data.length >= state.pagesize) {
                            state.loadingBol = true;
                        }
                    }).catch(err => {
                        // console.log(err)
                        reject(err)
                    })
                })
            },

            // 获取博客详情
            getBlogInfo({
                commit
            }, _id) {
                return new Promise((resolve, reject) => {
                    axios.get('blog/info', {
                        _id
```

```
            }).
            then(res => {
                commit('BLOGINFO', res)
                resolve(res)
            }).catch(err => {
                // console.log(err)
                reject(err)
            })
        })
    }
  }
}
export default blog
```

打开控制台可以看到对应的接口,如图 12.10 所示。

图 12.10　请求博客列表接口

12.5.3　博客详情页

博客详情页展示博客文章的完整内容。这里直接解析 markdown 内容并展示出来,代码示例如下:

```
<template>
  <article>
    <h1 class="title">{{blogInfo.title}}</h1>
    <div class="article-wrapper">
      <Back></Back>
      <div class="content">
        <div class="box">
          <Github          class="github"            :link="blogInfo.github" v-if="blogInfo.github"></Github>
          <div class="entry">
            <h1>{{blogInfo.title}}</h1>
```

```html
            <time>{{blogInfo.releaseTime | parseTime('{y}-{m}-{d}')}}</time>
            <div class="intro fmt" v-html="blogHtml"></div>
          </div>
          <div class="logo">
            <img
              :src="require('src/images/source_single_${blogInfo.source === 1?1:blogInfo.source === 2?2:3}.png')"
              :alt="blogInfo.title"
            />
          </div>
        </div>
      </div>
    </div>
  </article>
</template>

<script>
import { mapGetters } from "vuex";
export default {
  data() {
    return {
      blogHtml: ""
    };
  },
  async mounted() {
    await this.$store.dispatch("getBlogInfo", this.$route.params.id);
    this.blogHtml = this.blogInfo.html.replace(/<a /gi, `<a target='_blank'`);
    if (this.$route.query.type === "comment") {
      setTimeout(() => {
        this.goAnchor();
      }, 0);
    }
  },
  methods: {
    goAnchor() {
      let oComment = document.querySelector("#comment");
      let scrollTop = oComment.offsetTop;
      document.documentElement.scrollTop = scrollTop;
      document.body.scrollTop = scrollTop;
    }
  },
  computed: {
    ...mapGetters(["blogInfo"])
  },
  watch: {
    blogInfo() {
      document.title = this.blogInfo.title;
    }
  }
```

```less
};
</script>


<style lang="less" scoped>
.title {
  text-align: center;
  margin: 20px 0;
}
.article-wrapper {
  width: 7rem;
  max-width: 1000px;
  margin: 0 auto;
  padding: 20px;
  padding-top: 0;

  .content {
    width: 100%;
    padding: 10px;
    background: #f9f9f3 url("../../images/note-bg.jpg");
    text-shadow: 1px 1px 0 rgba(255, 255, 255, 0.25);

    .box {
      padding: 0.3rem 0.6rem;
      border: 1px dashed #c9c9c7;
      position: relative;
      .github {
        position: absolute;
        right: 0;
        top: 0;
      }

      .entry {
        line-height: 30px;
        h1 {
          margin-bottom: 20px;
          text-align: center;
          color: @theme-red-color;
        }
        time {
          color: #b2b2ae;
          font-size: 12px;
          margin-bottom: 20px;
          display: block;
          text-align: center;
        }
        .intro {
          overflow-x: scroll;
          font-size: 14px;
```

```css
      @media screen and (max-width: @pc-width) {
        font-size: 12px;
      }
    }
  }

  .logo {
    margin-top: 30px;
    margin-right: -20px;
    text-align: right;
    img {
      width: 50px;
    }
  }
}
</style>
```

单击博客列表中对应的博客，直接跳转到对应的博客详情，或者直接在浏览器中打开 http://localhost:8080/article/5df63796659b8fdd9f6865d9，显示的博客详情内容如图 12.11 所示。

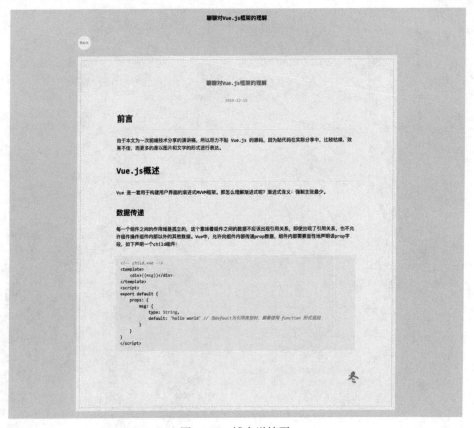

图 12.11　博客详情页

第 12 章 实战：基于 Koa+MongoDB 实现博客网站

切换到手机模式下，显示如图 12.12 所示。

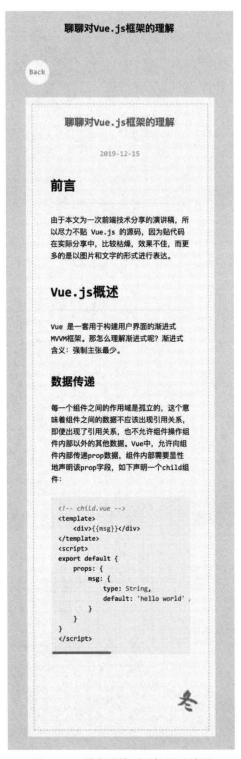

图 12.12 博客详情页手机展示效果